HACKER CULTURE

HACKER CULTURE

Douglas Thomas

University of Minnesota Press
Minneapolis
London

Published by the University of Minnesota Press
111 Third Avenue South, Suite 290
Minneapolis, MN 55401-2520
http://www.upress.umn.edu

Printed in the United States of America on acid-free paper

The University of Minnesota is an equal-opportunity educator and
employer.

Library of Congress Cataloging-in-Publication Data

Thomas, Douglas, 1966–
  Hacker culture / Douglas Thomas
       p. cm.
  ISBN 0-8166-3346-0
  1. Computer programming—Moral and ethical aspects. 2. Computer
hackers. I. Title.
QA76.9.M65 T465 2002
306.1-dc21
                                                                2001005377

13 12 11 10 09 08 07 06 05 04   10 9 8 7 6 5 4 3

Contents

Acknowledgments

This book was written with the help of a great number of people, many of whom I met only in passing, either at conventions and court hearings or in IRC chat and LISTSERV discussions. They contributed to the book in ways that are impossible to measure or account for here. Equally important are the people who have been a part of hacker culture or who have thought long and hard about it and took the time to share their insights. I especially want to thank Katie Hafner, John Perry Barlow, Jonathan Littman, John Markoff, Jericho, Mike Godwin, Wendy Grossman, Bruce Sterling, Chris Painter, David Schindler, Kevin Poulsen, Eric Corley, Don Randolph, Lewis Payne, Greg Vinson, Evian S. Sim, Michelle Wood, Kimberly Tracey, Mudge, The Deth Vegetable, Oxblood Ruffin, and many other members of the cDc and the L0pht for their help.

I also owe a debt of thanks to my colleagues at the Annenberg School for Communication, particularly Bill Dutton and Sandra Ball-Rokeach, who have been remarkably helpful and who have supported my work in countless ways. Lynn Spigel and Marsha Kinder both offered me help and insight in critical areas. My friends Peggy Kamuf, Philippa Levine, and Curt Aldstadt have offered their support, insight, and ideas throughout the writing of the book.

I am grateful to the Online Journalism Program at the University of Southern California, especially Larry Pryor and Joshua Fouts, who always gave me free rein to explore and write about hacking for the *Online Journalism Review*. The essays I wrote for the review allowed me to think through some difficult and complex questions, and their editorial style made that work much easier than I had any right to expect. James Glave, at *Wired News,* also provided an outlet for stories when they seemed to matter the most. The Southern California Studies Center helped fund a portion of the research through a junior faculty grant. For that I thank the center's director, Michael Dear.

The writing of this book was greatly assisted by Doug Armato and Will Murphy as well as by the comments from the reviewers at the University of Minnesota Press, who offered creative and constructive advice.

I benefited enormously from the careful readings offered by my friends Marita Sturken and Dana Polan. I thank them for their help and their friendship.

A special debt of gratitude goes to Kevin Mitnick and Chris Lamprecht, who shared their stories with me and helped me understand, in ways I had never imagined, what it meant to be a hacker in the 1990s.

Finally, I want to thank Ann Chisholm, who is the love of my life, who provides me with my inspiration, and who suggested, so long ago, that I might want to think about a "hacker project."

Introduction

Since the 1983 release of the movie *WarGames,* the figure of the computer hacker has been inextricably linked to the cultural, social, and political history of the computer. That history, however, is fraught with complexity and contradictions that involve mainstream media representations and cultural anxieties about technology. Moreover, hacking has its own history, which is itself as complex as it is interesting. In tracing out these intricate, intertwining narratives, this book is an effort to understand both who hackers are as well as how mainstream culture sees them. Part of the complexity is a result of the fact that these two constructions, hacker identity and mainstream representation, often 'reflect on each other, blurring the lines between fact and fiction.

The term "hacker" has its own historical trajectory, meaning different things to different generations.[1] Computer programmers from the 1950s and 1960s, who saw their work as breaking new ground by challenging old paradigms of computer science, think of hacking as an intellectual exercise that has little or nothing to do with the exploits of their 1980s and 1990s counterparts. Indeed, this older generation of hackers prefer to call their progeny "crackers" in order to differentiate themselves from what they perceive as their younger criminal counterparts. The younger generation take umbrage at such distinctions, arguing that today's hackers are doing the real work of exploration, made necessary by the earlier generation's selling out. In some ways, these younger hackers argue, they have managed to stay true to the most fundamental tenets of the original hacker ethic. Accordingly, the very definition of the term "hacker" is widely and fiercely disputed by both critics of and participants in the computer underground. Indeed, because the term is so highly contested, it gives a clue to both the significance and the mercurial nature of the subculture itself. Moreover, there seems to be little agreement within the academic literature on what constitutes hacking. In accounts that range from Andrew Ross's characterization of the hacker underground as "protocountercultural" to Slavoj Zizek's notion that "hackers operate as a circle of initiates who exclude themselves from everyday 'normality' to devote themselves to programming as an end in itself" to Sandy Stone's exposition of style at the Atari labs, whenever the complexity and intensity of technology are discussed, hackers are a primary cultural signifier.[2]

On top of the generational differences, hackers themselves, especially "new-school" hackers (of the 1980s and 1990s), have difficulty in defining exactly what hacking is. To some, it is about exploration, learning, and fascination with the inner workings of the technology that surrounds us; to others, it is more about playing childish pranks, such as rearranging someone's Web page or displaying pornographic images on a public server. It is, in all cases, undoubtedly about the movement of what can be defined as "boy culture" into the age of technology. Mastery over technology, independence, and confrontation with adult authority, traits that Anthony Rotundo has identified as constitutive of boy culture, all figure prominently in the construction of hacker culture. Even tropes of physical superiority and dominance have their part in the world of electronic expression. Such findings are hardly surprising, as the hacker demographic is composed primarily (but not exclusively) of white, suburban boys. There are relatively few girls who participate in the hacker underground, and those who do so oftentimes take on the values and engage in the activities of boy culture just as readily as their male counterparts.

While the "old-school" hackers were usually graduate students at large universities, their "new-school" counterparts are substantially younger, usually teenagers who have a particular affinity for technology. A primary reason for the difference in age has to do with access to and availability of technology. Where hackers of the 1950s, 1960s, and 1970s had little or no access to computers outside of a university environment, hackers in the 1980s and 1990s had access to the personal computer, which brought the technology that enabled hacking into their homes and schools. As a result, the newest generation of hackers have been able to work out a number of "boy" issues online, including the need to assert their independence and the testing of the boundaries of adult and parental authority. The introduction of the personal computer into the home, in the 1980s and 1990s, transformed a predominantly male, university culture into a suburban, youth culture and set these two histories, in part, against each other. The present work is an effort to situate and understand hacking as an activity that is conditioned as much by its history as by the technology that it engages. It is an effort to understand and at some level rethink the meaning of subculture in an electronic age, both through the means by which that subculture disputes meaning and makes meaning and through mass-mediated and cultural representations.

Hacking and Popular Culture

Because hacker subcultures flourished at the computer labs of MIT, Cornell, and Harvard in the 1960s and 1970s, they can be seen as constituting an institutional (even if occasionally resistant) culture. Hackers of the old school relied extensively on their institutions for support and access to machines. The problems that hackers solved in these university settings would eventually lead to the birth of the personal computer (PC) and launch an entire industry that would drive technological innovation. Most of these hackers would go on to form Silicon Valley start-up companies, lead the open-source software movement, and create small (or sometimes very large) fortunes for themselves. They entered the popular imagination not as hackers but as "computer geniuses" or "nerds."

Their progeny, the kids who would grow up with the PC in their homes and schools, were faced with a different set of problems and possibilities. These young hackers were born into a world of passwords and PIN numbers, created and made possible by the corporations that the old-school hackers had built. These younger hackers had no institutional affiliation and no limitations on access (at least to their own machines). Moreover, they saw that secrecy was a double-edged sword. Secrets can preserve an institution's identity, but, just as important, they can also protect a hacker from being identified. While a culture of secrecy provided for security, it also allowed for a new kind of anonymity, one that could be exploited and used to a hacker's advantage.

With these discoveries, the new-school hackers began to reach out to one another and create their own culture, a culture that expressed a general dissatisfaction with the world, typical of teenage angst, but also a dissatisfaction with ways technology was being used. For teenage boys discovering the ways that computers could be used to reach out to one another, there was nothing more disturbing than seeing those same computers being used to systematically organize the world. Groups of hackers began to meet, to learn from one another, and to form a subculture, which was dedicated to resisting and interrupting "the system."

As the underground was developing into a bona fide subculture, popular culture was not letting the hacker phenomenon go unnoticed. In the early 1980s a new genre of science fiction literature emerged that began to color the underground's ethos. It, and particularly the work of William Gibson, was the literature of cyberpunk which would give hackers a set of heroes (or antiheroes) to emulate. The world of cyberpunk portrayed a high-tech outlaw culture, where the rules were made up by those on the frontier — not by bureaucrats. It was a digital world, where the only factor that mattered was how smart and talented you were. It was in this milieu that Gibson would coin the term "cyberspace":

> Cyberspace. A consensual hallucination experienced daily by billions of legitimate operators, in every nation, by children being taught mathematical concepts.... A graphic representation of data abstracted from the banks of every computer in the human system. Unthinkable complexity. Lines of light ranged in the nonspace of the mind, clusters and constellations of data. Like city lights, receding... [3]

Gibson called those who roamed this space "console cowboys," data jockeys who could manipulate the system toward the ends of digital espionage or personal gain. These hackers believed they were describing a future where they would feel at home, even if that home was a dystopia where the battle over information had been fought and lost, a world of what Thomas M. Disch calls "pop despair," in which the dystopian view of the future is "ameliorated only by two elements: fashion and an interior life lived in cyberspace."[4] What is intriguing about Gibson's characters is not that they exist in this world, but that they don't seem to mind it. Gibson's ne'er-do-well protagonists completely accept the world they inhabit. They do not protest or even desire to see things differently. Instead, they inhabit and rule a world in which they exercise near-complete control. As Bruce Sterling points out, it is the ideal model for disaffected suburban youth culture.[5] Where the suburban landscape provides little of interest for youth culture, the world of computers and networks provides a nearly infinite world for exploration.[6] The typical hacker is a white, suburban, middle-class boy, most likely in high school. He is also very likely self-motivated, technologically proficient, and easily bored. In the 1980s and even the 1990s, computers became a tool for these youths to alleviate their boredom and explore a world that provided both an intellectual challenge and excitement. But it was also a world that was forbidden — a world of predominantly male authority into which they could trespass with relative ease, where they could explore and play pranks, particularly with large institutional bodies such as the phone companies. It was a world of excitement that allowed them to escape the home and be precisely the "noise" in the system that they had fantasized about.

It would take nearly a decade for mainstream culture to catch up with the hacker imagination. In 1989, Clifford Stoll wrote *The Cuckoo's Egg*, a tale of international espionage that detailed his manhunt for hackers who had broken into U.S. military computers and had spied for the KGB. Stoll's tale was part high-tech whodunit, part cautionary tale, and all high drama. *The Cuckoo's Egg* stayed on the *New York Times* bestseller list for four months. Soon after, Katie Hafner and John Markoff published *Cyberpunk,* which told stories about three hackers — Robert Morris, Kevin Mitnick, and Pengo, each of whom achieved notoriety for hacking exploits ranging from crashing computer systems to international espionage. Stoll's and Hafner and Markoff's books captured the national imagination and portrayed hackers in highly dramatic narratives, each of which ended with the hacker's capture, arrest, and prosecution.

With the publication of these two books, the image of the hacker became inextricably linked to criminality.

Fear was driving the popular imagination, and hackers were delighted to go along with the image. After all, what high school kid doesn't delight in the feeling that he or she rules a universe that their parents, teachers, and most adults don't understand? One thing teenagers understand is how to make their parents uncomfortable. Like loud music, teen fashion, and smoking cigarettes, hacking is a form of rebellion and an exercise of power. The difference rests in the fact that the 1990s represented such a fundamental break between youth and mainstream culture that hacking was unable to be successfully assimilated into the narratives of youth rebellion without being either wildly exaggerated or completely trivialized. Parents intuitively understand the defiance of music, youth fashion, and cigarettes; they did similar things themselves. With hacking, they are faced with an entirely new phenomenon. That gap, between what hackers understand about computers and what their parents don't understand, and more importantly fear, makes hacking the ideal tool for youth culture's expression of the chasm between generations. Hacking is a space in which youth, particularly boys, can demonstrate mastery and autonomy and challenge the conventions of parental and societal authority. Divorced from parental or institutional authority, the PC enabled the single most important aspect of formative masculinity to emerge, independent learning, "without the help of caring adults, with limited assistance from other boys, and without any significant emotional support."[7] Hackers used the personal computer to enter the adult world on their own terms. In doing so, they found a kind of independence that was uniquely situated. Hackers had found something they could master, and unlike the usual rebellious expressions of youth culture, it was something that had a profound impact on the adult world.

The 1980s and 1990s also saw the productions of a several films that had hackers as primary figures, further imbedding their status as cultural icons. In 1982, *TRON* captured the public imagination with the vision of the ultimate old-school hacker, who creates, according to Scott Bukatman, "a phenomenological interface between human subject and terminal space," a literal fusion of the programmer and the computer, the ultimate cyberpunk fantasy.[8] More recently, *Pi* (1999) provided a dark mirror of the cyberpunk vision, where the hacker is driven mad by his obsession with technology and its ability to decipher nature and the world. In other narratives, hackers often served as technologically savvy protagonists. In films like *Sneakers* (1992), *The Net* (1995), and *The Matrix* (1999), hackers serve as central figures who are able to outwit the forces of evil based on an extraordinary relationship to technology. Television presented a similar view — the "lone gunmen" on the *X-Files* and series such as *The Net, VR5,* and *Harsh Realm* all presented hackers as technologically sophisticated protagonists able to perform acts of high-tech wizardry in the service of law enforcement or the state.

Although the figure of the hacker was widespread in media representation, two films in particular influenced the hacker underground and, to a large degree, media representation of it. Those films, *WarGames* (1983) and *Hackers* (1995), had a disproportionate influence on hacker culture, creating two generations of hackers and providing them with cultural touchstones that would be, at least in part, the basis for their understanding of hacking. While films like *Sneakers* and *The Net* are of great interest to hackers, they are often evaluated based on their factual accuracy or technical sophistication, rather than as cultural touchstones for hacker culture. A primary difference is the opportunities for identification that each film provides. While *WarGames* and *Hackers* had male, teenage protagonists, *Sneakers* and *The Net* provided barriers to identification: in *Sneakers*, Robert Redford's character was in his late forties or even early fifties, according to the chronology of the film's narrative, and *The Net* starred Sandra Bullock, presenting a female protagonist whom teenage boys were more likely to see as an object of desire than of identification.

As a result, hackers, when discussing films that interest them, are much more likely to speak of *WarGames* and *Hackers* in terms of influence, while referring to films such as *The Net, Sneakers,* and *Johnny Mnemonic* in terms of how much they liked or disliked the film or whether or not it accurately represented hackers and technology itself.

Hacking as Boy Culture

Boy culture has a number of historically situated ideals and values that have been put into play in the history of hacking, both by the youngest generation of hackers and by their older, university counterparts. Perhaps the most important element of hacker culture is the notion of mastery. As Rotundo argues, this element is complex and involves "constantly learning to master new skills," as well as mastering one's social and physical environment.[9] It is also a culture of competition, where affection is expressed through "playful spontaneity," "friendly play," and "rough hostility," whereby boys learn to express "affection through mayhem."[10] In earlier manifestations of boy culture, that affection was shown through physical contact, contests, and "roughhousing," activities that provided physical contact under the cover of aggression. With hackers, such contests continue but are marked by a technological transformation. The absence of the body makes physical contact impossible. Such contact is replaced by tropes of emotional aggression and ownership. Hackers commonly taunt each other with threatening overtures designed to provoke fear (or, as they more commonly spell it, "phear," in a language marked by technology, in this case the phone), publicly challenge each other's knowledge, and routinely accuse others of being less skilled, less knowledgeable, or "lame." The goal of the aggression is complete domination over another hacker (or other target), expressed through the notion of ownership. In hacker terms, phrases such as "r00t owns you" or "I'll own your ass" express both mastery and subordination. They express a fantasy of complete technological domination and control over others, the idea that the vanquished hacker (or system) is at the mercy of the more powerful and skilled hacker. Even though the hackers of the 1990s (and to a lesser degree of the 1980s) enact this kind of aggression, typical of boy culture, they also share a number of qualities with their older, university counterparts.

The traits most strongly shared by the two generations of hackers are the desire for mastery over technology and the struggle between authority and autonomy that constitutes a significant portion of formative masculinity and youth culture in contemporary society.[11]

While both generations of hackers enact these values, they do so in different ways. For hackers in the university context, a premium was placed on absolute mastery over the machine; hacking was seen as a way of life; and battles over autonomy and authority took place within the confines of the institution, usually between the hackers and their professors or university administrators.[12] As the PC entered the home in the 1980s, however, hacking became a viable means for groups of predominantly white, teenage boys to create a space for their own youth culture. These new-school hackers saw technology as a means to master both the physical machines and the social relations that were occurring through the incorporation of technology into everyday life (such as ATM machines, institutional records becoming computerized, the growth of the Internet, and so on). Their control over computers, they realized, was an ideal vehicle for teenage boy mayhem. It was also a tool for testing and reinforcing boundaries. The computer was in the home but was also a connection to a world outside the home. It could touch the world in playful, mischievous, and even malicious ways.

Technology and the Postmodern Turn

While hacking and hackers can be easily positioned in terms of youth culture and the culture of secrecy, hacking also has a broader social and cultural set of implications for how we look at the world.[13] The medium of the computer affords a particular avenue of resistance that speaks to broader questions of technology and culture. In terms of these broader questions, two specific issues arise. First, hacker culture is a "postmodern" moment that defines a period in which production is being transformed from a stable, material, physical system to a more fluid, rapid system of knowledge production.[14] For example, the emergence of software as a codified form of knowledge is caught between and negotiated by these two poles. Software is merely an arrangement of bits, stored on a medium, making it an ideal example of a knowledge-based system of production.[15] When reduced to its most basic parts, all software is nothing more than a series of 0s and 1s; the medium of distribution is wholly irrelevant to the content (unlike other forms of mediated knowledge). It is merely the translation of thought into a codified and distributable form. But at present, the material system of production impinges on that knowledge directly. The distribution of software (which requires little more than making it accessible on the Internet) is hindered by the physical processes characteristic of earlier modes of production: copying the software, putting it on disks or CD-ROMs, packaging the software, shipping it to retail outlets, and so on. In short, software is sold as if it were hardware. Knowledge, which has always needed to be commodified into some material form (for example, books), now can be transmitted virtually without any material conditions at all. For the first time, we are seeing an actualization of the basic principle that knowledge is virtual. What hackers explore is the means by which we are beginning to redefine "knowledge in computerized societies."[16] In this sense, hackers can help us understand the manner in which culture is both resistant to the transformation of knowledge and inevitably shaped by it. I also argue that hackers help us understand the transformations taking place around us not only through their analysis of and reaction to them but also by the manner in which they are represented in mainstream media and culture.

The second issue has to do with postmodernity's relationship to the body and identity, two themes that I argue are at the heart of the intersection between hacker and mainstream culture. One of the primary means by which modern culture has been questioned and destabilized by postmodernity is through a radical questioning of the idea of a stable identity. Much of the postmodern critique centers on the idea that neither the body nor "identity" can be seen as a stable or unified whole.[17] Instead, identity is composed in a fragmentary manner, suggesting that it is both more fluid and more complex than had been previously theorized. Postmodernism also questions the manner in which the body has been utilized to construct stable positions of identity (such as sex, gender, race, class, and so on). Such challenges disrupt the sense of certainty that characterizes modernity. Accordingly, theories of postmodernism provide an ideal tool to examine hacker culture in the sense that the hacker underground targets and exploits stable notions of identity and the body in its hacking activities (for example, the idea that knowing a secret such as a password can confirm the physical identity of a person). As such, hacking becomes more than a simple exercise of computer intrusion; instead, in this broader context, it enacts a challenge to a host of cultural assumptions about the stability of certain categories and cultural norms regarding identity and the body. In most cases, hackers are successful because they are able to play upon assumptions about stability of identity and bodies while actively exploiting precisely how fluid and fragmented they actually are.

Hacking Culture

In writing this book, I have often found myself at the nexus of several positions: between ethnographer and participant, between academic and advocate, between historian and storyteller. It became apparent to me very early on that it would be impossible to divorce my own personal experience and history from this book and that to attempt to do so would make for an overly cautious book.

What I attempt to offer here is part genealogy, part ethnography, and part personal and theoretical reflection. As a genealogy, this book is an effort to produce what Michel Foucault referred to as "local criticism," as criticism that "is an autonomous, non-centralised kind of theoretical production, one that is to say whose validity is not dependent on the approval of the established regimes of thought."[18] Local criticism takes as its focus the insurrection of subjugated knowledges. These are knowledges that have been buried and disguised and that through examination allow us to examine the ruptures and fissures in what is assumed to be a coherent and systematic regime of thought, history, or theory. Indeed, local criticism is an effort to recover precisely those ideas that have either been excluded, forgotten, or masked in the process of creating historical narratives.

Those ideas are also a kind of "popular knowledge," which is not meant in the sense of "popular culture," but, rather, is defined as being a differential knowledge that cannot be integrated or uniformly woven into a single narrative. Its force is generated by the very fact that it opposes "the conventional narratives that surround it."[19] Taking such a perspective enacts what Foucault defined as *genealogy,* as the "union of erudite knowledge and local memories which allows us to establish a historical knowledge of struggles and to make use of this knowledge tactically today."[20] In the case of hacker culture, these two knowledges, both buried and popular, are found in the discourses of the underground itself, on the one hand, and of the media, popular culture, and law, on the other hand. As a result, this work has been pulled in two directions at once — first, toward the erudition and excavation of buried and subjugated knowledges that the study and examination of the discourse *about* hackers demand, and, second, toward the more ethnographic and personal research that is required to understand the discourse *of* hackers.

This work explores the "computer underground" through an examination of the subculture of hackers and through an understanding of hackers' relationship to mainstream contemporary culture, media, and law. In particular, I argue that hackers actively constitute themselves as a subculture through the performance of technology.

By contrast, I contend that representations of hackers in the media, law, and popular culture tell us more about contemporary cultural attitudes about and anxiety over technology than they do about the culture of hackers or the activity of hacking. Although these media representations of hackers provide an insight into contemporary concerns about technology, they serve to conceal a more sophisticated subculture formed by hackers themselves. Through an examination of the history of hacking and representations of hackers in film, television, and journalistic accounts, and through readings of key texts of the hacker underground, I detail the ways in which both the discourse about hackers and the discourse of hackers have a great deal to tell us about how technology impacts contemporary culture.

Hacker subculture has a tendency to exploit cultural attitudes toward technology. Aware of the manner in which it is represented, hacker culture is both an embracing and a perversion of the media portrayals of it. Hackers both adopt and alter the popular image of the computer underground and, in so doing, position themselves as *ambivalent* and often *undecidable* figures within the discourse of technology.

In tracing out these two dimensions, anxiety about technology and hacker subculture itself, I argue that we must regard technology as a *cultural* and *relational* phenomenon. Doing so, I divorce the question of technology from its instrumental, technical, or scientific grounding. In fact, I will demonstrate that tools such as telephones, modems, and even computers are incidental to the actual *technology of hacking*. Instead, throughout this work, I argue that what hackers and the discourse about hackers reveal is that technology is primarily about mediating human relationships, and that process of mediation, since the end of World War II, has grown increasingly complex. Hacking, first and foremost, is about understanding (and exploiting) those relationships.

Accordingly, the goal of this work is one that might be called "strategic," in Foucault's sense of the word, an intervention into the discourse of hackers and hacking that attempts to bring to light issues that have shaped that discourse. Therefore, this book positions hackers and hacker culture within a broader question of the culture of secrecy that has evolved since the 1950s in the United States. Hackers, I contend, can help us better understand the implications of that aspect of secrecy in culture. Conversely, the emerging culture of secrecy can help us better understand hackers and hacker culture.

In the past twenty years, the culture of secrecy, which governs a significant portion of social, cultural, and particularly economic interaction, has played a lead role in making hacking possible. It has produced a climate in which contemporary hackers feel both alienated and advantaged. Although hackers philosophically oppose secrecy, they also self-consciously exploit it as their modus operandi, further complicating their ambivalent status in relation to technology and contemporary culture. The present project explores the themes of secrecy and anxiety in relation to both contemporary attitudes toward technology and the manner in which hackers negotiate their own subculture and identity in the face of such cultural mores.

The book begins by examining the culture of secrecy and the basic representation of a hacker with which most readers will be familiar — the high-tech computer criminal, electronically breaking and entering into a bank using only a computer and a phone line. This representation is problematized through a repositioning of hacking as a cultural, rather than technical, activity. The old-school hackers of the 1960s and 1970s — who are generally credited with the birth of the computer revolution and who subscribed to an ethic of "free access to technology" and a free and open exchange of information — are thought to differ from their 1980s and 1990s counterparts, generally stereotyped as "high-tech hoodlums" or computer terrorists. Historically, however, the two groups are linked in a number of ways, not the least of which is the fact that the hackers of the 1980s and 1990s have taken up the old-school ethic, demanding free access to information. Further problematizing the dichotomy is the fact that many old-school hackers have become Silicon Valley industry giants, and, to the new-school hackers' mind-set, have become rich by betraying their own principles of openness, freedom, and exchange. Accordingly, the new-school hackers see themselves as upholding the original old-school ethic and find themselves in conflict with many old schoolers now turned corporate.

Overview

In the 1980s, hackers entered the public imagination in the form of David Lightman, the protagonist in the hacker thriller *WarGames* (1983), who would inspire a whole generation of youths to become hackers, and later, in 1988, in the form of Robert Morris, an old-school hacker who unleashed the Internet worm, bringing the entire network to a standstill. These two figures would have significant influence in shaping hacker culture and popular media representations of it. From the wake of these public spectacles would emerge the "new school," a generation of youths who would be positioned as heroes (like Lightman in *WarGames*) and villains (like Morris) and who, unlike the old-school hackers two decades earlier, would find little or no institutional or government support.

The new school emerged in an atmosphere of ambivalence, where hacking and hackers had been seen and celebrated both as the origins of the new computerized world and as the greatest threat to it. New-school hackers responded by constituting a culture around questions of technology, to better understand prevailing cultural attitudes toward technology and to examine their own relationship to it as well.

This book traces out the history and origins of hacker culture in relation to mainstream culture, the computer industry, and the media. Chapter 1 introduces the basic questions that motivate this study: Will hackers of the new millennium exert the same level of influence on the computer industry's new pressing concerns (such as privacy, encryption, and security) as the old-school hackers of the 1960s did in their time by creating the industry itself? Are hackers still the central driving force behind innovation and design in the rapidly changing computer industry? And what effects will the children of the 1980s, raised on *WarGames* and the legend of Robert Morris, have on the future of computing as they too leave college, run systems of their own, and take jobs with computer companies, security firms, and as programmers, engineers, and system designers?

Chapter 2 explores the manner in which hacker culture negotiates technology, culture, and subculture, beginning with a discussion of the relationship between hackers and technology. Hackers' discourse demonstrates the manner in which they perceive technology as a "revealing" of the essence of human relationships. In this sense, technology reveals how humans are ordered by the technology that they use. Or, put differently, hackers understand the ways in which technology *reveals* how people have been defined by technology. In detailing the implications of this argument, I analyze the various ways that hackers use language, particularly through substitutions and transformations of different letters to mark the technology of writing as something *specifically* technological. In particular, I examine the manner in which particular letters and numbers, which are often substituted for one another, can be traced to specific technologies (such as the keyboard) as a reflexive commentary on the nature and transformations of writing in relation to technology itself. An example of this kind of writing would be the substitution of the plus sign (+) for the letter *t,* the number 1 for the letter *l,* and the number 3 for the letter *E.* In one of the more common examples, the word "elite" becomes 3l33+.

I also examine the manner in which hackers exploit their understanding of human relationships. One such strategy, referred to by hackers as "social engineering," is a process of asking questions or posing as figures in authority to trick people into giving them access or telling them secrets.

This sense of revealing is further reflected in the text "The Conscience of a Hacker" (often called "The Hacker Manifesto"). Written by The Mentor and published in *Phrack* shortly after its author's arrest, it is an eloquent description of, among other things, how hackers regard contemporary society and its relationship to technology. The essay is widely cited and quoted and appears on Web pages, on T-shirts, and in films about hackers. In my reading of the document, I argue that it illustrates a sophisticated understanding of the relationship between hacking and culture, particularly in relation to issues of performance.

The third chapter explores the new directions that hacking has taken in the past decade. The growth of the Internet, the development of PC hardware, and the availability of free, UNIX-like operating systems have all led to significant changes in hackers' attitudes toward technology.

The growth of LINUX, a free clone of the UNIX operating system, in the past five years has made it possible for hackers to run sophisticated operating-system software on their home computers and, thereby, has reduced the need for them to explore (illegally) other people's systems. It has made it possible, as well, for anyone with a PC and an Internet connection to provide a web server, e-mail, access to files, and even the means to hack their own systems.

Most important, however, has been the relationship between hackers and the computer industry. As many hackers from the 1980s have grown older, they have entered the computer industry not as software moguls or Silicon Valley entrepreneurs but as system managers and security consultants. Accordingly, they still find themselves in conflict with the industry. Many hackers see themselves in the role of self-appointed watchdogs, overseeing the computer industry and ensuring that the industry's security measures meet the highest standards. These hackers are not above either shaming companies into making changes or even forcing improvements by publicly releasing files that "crack" the security of what they perceive to be inferior products.

Hackers have also become more social in the late 1980s, the

1990s, and the first years of the twenty-first century. Previously confined to "electronic contact," hackers have started organizing conventions, or "cons," which feature speakers, vendors, events such as "Hacker Jeopardy," and even a "spot the Fed" contest in which hackers can win T-shirts for identifying federal agents. These conventions have become an important means of sharing and disseminating information and culture.

As hacking events have become more social, hackers have started to band together around areas of common concern, and such gatherings have begun to politicize hacker culture in the process. As battles over the restriction of the export and creation of encryption become more focused, hackers are finding new issues to explore. In relation to secrecy, hackers are working to find better and faster ways to break encryption routines, but of equal concern is the manner in which encryption is being put to use. At the most radical level, one of the oldest (and most flamboyant) groups in the computer underground, the Cult of the Dead Cow, has undertaken a project of supplying a group of Chinese dissidents, who call themselves the Hong Kong Blondes, with encryption software to secure their communications against government eavesdropping and with computer intrusion techniques to perform acts of resistance in response to human rights violations.

As the corporate computer industry, the government, and hackers come into increasing contact, the hacker underground is continually reshaping itself as a response to those relationships. General principles regarding the nature and role of technology continue to shape a hacker ethic, which promises to reinvent itself with each new set of developments and each shift in cultural attitudes toward and anxieties about technology.

An examination of hackers' relationships to technology reveals the ways in which technology serves as both the basis for the constitution of their own subculture and the point of division from mainstream culture. That distinction is further reflected in chapters 4 and 5, which analyze the long-standing journal *Phrack* and the ways in which hackers deploy technology as resistance. These chapters continue the examination of hacker culture through a reading of the online journal *Phrack,* a journal written by and for the computer underground, and through a comparative analysis of a hacker video and the MGM film *Hackers* (1995).

In my reading of *Phrack,* I explore the ways in which hacker culture is both understood and disseminated through the creation of an *electronic sense of style.* As a subculture, the hacker underground is reflected by the various meanings and styles represented in *Phrack,* not only in its articles and features but also in its interaction with the broader social culture. To that end, I examine two cases: first, the prosecution of a *Phrack* editor for publishing a supposedly secret BellSouth 911 document, copied by a hacker from a BellSouth computer system; and, second, *Phrack*'s institution of a copyright statement that rendered the magazine free to hackers but forced law enforcement and corporations to subscribe. In each case, what is revealed is *Phrack*'s relationship to the social, cultural, and political dimensions of technology and the hacker underground's negotiation with contemporary culture.

The comparison of a video made by hackers (documenting the breaking into and subsequent hacking of a telephone control office) and the film *Hackers* illustrates the manner in which hacker style is both defined by hackers and incorporated by mainstream culture. Each film marks a particular take on the phenomenon of hacking, and while the events in each are often parallel, the style illustrates a marked difference. The chapter concludes with an examination of the efforts of hackers to hack the MGM Web page announcing the release of the film as an act of resistance and as an effort to resist the incorporation of hacker style.

Chapter 6 comes backs to the question of the representation of hackers in popular and juridical discourse, focusing on depictions of hackers that emphasize criminality. I develop the idea that discourse about hackers' criminality is centered on issues of the body, addiction, and performance. In particular, I flesh out these themes by reading the cases of specific hackers (Kevin Mitnick, Kevin Lee Poulsen, and members of the Legion of Doom and the Masters of Deception) who were "tracked" and, ultimately, captured and jailed.

The book concludes by examining the cases of two hackers: Chris Lamprecht (aka Minor Threat), the first hacker to be banned from the Internet, and Kevin Mitnick, who was convicted on a twenty-five-count federal indictment for copying proprietary software from a cellular phone manufacturer and who has been the subject of numerous books, articles, and an upcoming feature film.

By tracing out hacker culture, from its origins in the 1950s and 1960s through the various transformations it has taken in the 1980s and 1990s, *Hacker Culture* marks the various and complex ways in which technology has played a pivotal role in the formulation of the hacker underground and in the public, popular, and legal representation of it. Marking such transformations not only provides a sense of where hacker culture has come from but also comments on the role of technology in mainstream culture and illustrates the ways in which technology has been woven into the fabric of American society. Over the next decade, we can expect to see changes in the roles that hackers take on, the manner in which they negotiate their identity, and the ways in which they inform culture about the role of technology in the practice of everyday life.

Part I

The Evolution
of the Hacker

The Evolution of the Hacker

The term "hacker" has been stretched and applied to so many different groups of people that it has become impossible to say precisely
what a hacker is. Even hackers themselves have trouble coming up
with a definition that is satisfactory, usually falling back on broad
generalizations about knowledge, curiosity, and the desire to grasp
how things work. For the purposes of this book, I want to think of
hackers as a culture, a group of computer enthusiasts who operate in
a space and manner that can be rightly defined by a sense of boundless curiosity and a desire to know how things work, but with the
understanding that such knowledge is further defined by a broader
cultural notion: secrecy. To understand today's hackers, it is essential
to understand the history of computers and computer culture.

The history of the computer is also a history of secrecy and of the
machines that have made those secrets possible and even necessary.
Hackers, since the beginning, have been a part of that story, and
while it is important to point out that the first generation of hackers
included the "father of the PC" and the creators of the Internet, they
were also the architects of a revolution in secrecy that has made the
PIN number an essential part of our daily existence.

The hackers I discuss in what follows are caught halfway between
these two worlds. They acknowledge, but refuse to accept, the manner in which secrecy has become a part of our daily routine. They
are interested in the ways that technology and secrecy interface and
how that combination can be explored, exploited, and manipulated
to their own advantage. To that end, hacker subculture is defined
more by an ethos than by technological sophistication or the ability
to program computers. Instead, today's hacker culture is defined in
large measure by the way it responds to mainstream culture, particularly in terms of issues of secrecy. Hacker culture, however, has a
dual nature. On the one hand, hackers have a culture that is very

3

much their own. They have their own norms, terminology, conferences, meeting places, and rules for conduct. On the other hand, their
culture is wholly dependent on mainstream culture and not merely
as something to react against, but rather as grounds for exploration.
The hacker ethos is self-generated, but the substance, the content of
hacker culture, is derived from mainstream culture's embrace of and,
simultaneously, confusion about technology. It is a culture in one
way completely divorced from mainstream culture, yet in another
way completely dependent upon it.

In order to understand who hackers are and what they do, it is
first necessary to trace their history, which follows a trail from the
computer labs of Harvard, MIT, and Cornell in the 1960s to the halls
of Redmond, Washington, the home of Microsoft, in the 1990s. But
hackers cannot be understood solely in terms of the technology with
which they are intertwined. Hackers and hacking are much more
about a set of social and cultural relations, which involves not only
the ways in which hackers exploit those social relations but also the
ways in which the image of the hacker has been created, refined, and
used as a symbol in popular culture to understand technology and
to give a face or image to the fears, uncertainties, and doubts that
accompany technological change.

Chapter 1

Hacking Culture

Fry Guy watched the computer screen as the cursor blinked. Beside him a small electronic box chattered through a call routine, the numbers clicking audibly as each of the eleven digits of the phone number was dialed. Then the box made a shrill, electronic whistle, which meant the call had gone through; Fry Guy's computer . . . had just broken into one of the most secure computer systems in the United States, one which held the credit histories of millions of American citizens.
— Paul Mungo and Bryan Clough, *Approaching Zero*

This is the common perception of today's hacker — a wily computer criminal calling up a bank or credit card company and utilizing mysterious tools to penetrate simply and effortlessly the secure system networks that hold our most important secrets. However, any attempt to understand today's hackers or hacking that only examines the blinking cursors and whistling boxes of computing is destined to fail. The reason is simple: hacking has never been just a *technical* activity. Perhaps the most striking illustration of this is that William Gibson, who in his book *Neuromancer* coined the term "cyberspace" and who invented a world in which hackers feel at home, for nearly a decade refused to have an e-mail address. In fact, *Neuromancer,* the book that has been (rightly or wrongly) held accountable for birth of the "new breed" of hackers, is rumored to have been written on a manual typewriter.[1] Hacking, as Gibson's work demonstrates, is more about the imagination, the creative uses of technology, and our ability to comment on culture than about any tool, computer, or mechanism. The hacker imagination, like the literature that it is akin to, is rooted in something much deeper than microchips, phone lines, and keyboards.

The current image of the hacker blends high-tech wizardry and criminality. Seen as the source of many evils of high-tech computing,

5

from computer espionage and breaking and entering to the creation of computer viruses, hackers have been portrayed as the dangerous other of the computer revolution. Portrayals in the media have done little to contradict that image, often reducing hackers to lonely, malicious criminals who delight in destroying sensitive computer data or causing nationwide system crashes.

In both the media and the popular imagination, hackers are often framed as criminals. As Mungo and Clough describe them, hackers are members of an "underworld" who "prowl through computer systems looking for information, data, links to other webs and credit card numbers." Moreover, hackers, they argue, can be "vindictive," creating viruses, for instance, that "serve no useful purpose: they simply cripple computer systems and destroy data. . . . In a very short time it has become a major threat to the technology-dependent societies of the Western industrial world."[2]

At least part of the reason for this impression rests with the media sensations caused by a few select cases. Clifford Stoll, as documented in his book *The Cuckoo's Egg,* for example, did trace a European hacker through the University of California at Berkeley's computer systems, ultimately revealing an attempt at international espionage, and the cases of Kevin Mitnick and Kevin Lee Poulsen, two hackers who were both arrested, prosecuted, and sent to prison for their hacking, gained considerable attention in the media and in subsequent books published about their exploits. Most of these accounts are journalistic in style and content and are more concerned with describing the events that took place than with analyzing the broader context out of which hackers have emerged. While the image of the hacker as a "criminal" seems to have taken over in the popular imagination, the broader context of the computer "underground" and, most important, the historical context force us to question such an easy categorization of this complex and varied subculture.

Between Technology and the Technical: Hacking as a Cultural Phenomenon

In the 1980s and 1990s, hackers were the subject of numerous films, TV shows, and news reports, most of which focused on the connection between hacking and criminality. As Joe Chidley describes them in an article for *Maclean's Magazine,* "Hackers are people who simply love playing with computers," but there is a "malicious subset of the hacker community, who intrude on computer networks to do damage, commit fraud, or steal data," and these hackers "now have an arsenal of technologies to help them in their quest for secrets."[3]

Within such a limited framework, which reduces hackers to criminals with their "arsenal of technologies," it makes little sense to speak of a "culture" of hacking. Hacking appears to be, like most crime, something that malicious people do for reasons that don't always seem to make sense. Why would a talented computer programmer choose to write a virus rather than write a program that might be more useful and, potentially, economically more rewarding? Why would hackers break into unknown systems when their talents could be employed in many other more productive ways? These questions make the hacker's goals and motivations difficult to decipher.

Rather than attempting to understand the motivation behind hacking, the media and computer industry instead focus on the manner in which computers are hacked. At this level, hackers are easy to understand — they have a specialized set of tools, and they use those tools to commit crimes. This basic theme was central to the protest against the release of SATAN (Security Administrator Tool for Analyzing Networks), a network-analysis tool that tests systems for security flaws. The program, which was written to make system administrators aware of security flaws that were already well known and often exploited by hackers, was publicly released by its authors, Dan Farmer and Wieste Venema, in April 1995. The release was met with an outpouring of anxiety about the future of Net security and fear that the public availability of the tool would turn average computer users into criminals. As the *Los Angeles Times* remarked, "SATAN is like a gun, and this is like handing a gun to a 12-year-old."[4] Other newspapers followed suit, similarly invoking metaphors of increasing firepower — "It's like randomly mailing automatic rifles to 5,000 addresses. I hope some crazy teen doesn't get a hold of one," wrote the *Oakland Tribune,* only to be outdone by the *San Jose Mercury News*'s characterization: "It's like distributing high-powered rocket launchers throughout the world, free of charge, available at your local library or school, and inviting people to try them out by shooting at somebody."[5] The computer industry was more sober in its analysis. "The real dangers of SATAN," as one advisory argued, "arise from its ease of use — an automated tool makes it very easy to probe around on the network."[6]

The basic objection to the release of SATAN was that it provided a tool that made system intrusion too easy, and making the program publicly available prompted outcries from those afraid that anyone with the tools could (and would) now invade systems. Omitted from most stories was the fact that SATAN had been available in a less powerful form as freeware (a freely distributed software package, accessible to anyone) on the Internet for years, along with several other programs that provided similar functions, not to mention a host of more powerful programs that were already widely available for the express purpose of unauthorized system entry. Additionally, SATAN only tested computer systems for well-known, previously discovered (and easily fixed) security holes.

SATAN was nothing new, but the discussion of it was. This response illustrated how convinced the general public was that the threat of hacking rested in the tools. While the apocalyptic effects of SATAN's release failed to materialize (no significant increase in any system intrusion has been reported, nor has any been attributed to SATAN since its release), the anxieties that SATAN tapped into are still present. The response to SATAN was in actuality a response to something deeper. It was a reaction to a basic cultural anxiety about both the complexity of technology and the contemporary culture's reliance upon that technology. SATAN appears to give anyone who wants it the tools to disrupt the system that very few people understand yet that everyone has come to rely on in their daily lives.

A cursory examination of both public and state responses reveals a paranoia regarding the hacker that one can easily attribute to a Luddite mentality, a generation gap, or a pure and simple technophobia that seems to pervade U.S. culture. While these aspects are a very real part of contemporary culture, such a simple set of answers covers over more than it reveals. Most of the responses to hackers and hacking have served to lower the level of public discussion by confusing hackers with the tools that they use and making hyperbolic equations between computer software and high-power munitions. Like any other social and cultural phenomenon, the reasons for the growth of hacking in the United States (and as an international phenomenon) are myriad, and the reactions to hacking often reflect a wide range of reactions, from hope and fear to humor and dismay.

The responses to hacking — in the popular imagination and in the minds of agents of law enforcement and the criminal justice system, a response documented in court records, TV shows, movies, newspapers, books, and even Web pages — reveal more about contemporary culture than about hackers and hacking. However, much as was the case with SATAN, public reaction to hackers both tells us a great deal about the public that is reacting and, ironically, shields us from an understanding of the complexities and subtleties of the culture of the computer underground. By simply equating hackers with the tools they use, the media and popular representations of hackers have failed to understand or account for even the most basic motivations that pervade hacker culture.

In trying to determine what hacking is and what hacker culture looks like, I make a distinction between technology, as a broad, relational, and cultural phenomenon, and the technical or scientific, the products of technology itself (for example, telephones, computers, and modems).[7] In doing so, I am also separating hackers' culture and motivation, which are very much about technology, from the idea of tools or specific technical items, which are for the most part incidental to the idea of hacking. These two concepts, technology and the technical, are different in kind, and to understand what constitutes hacking, we need to be careful to examine these two ideas as separate entities. Technology should be considered a *cultural* phenomenon, and in that sense, it tells us primarily about human relationships and the manner in which those relationships are mediated. The technical, by contrast, is concerned only with the instrumental means by which those relationships occur. It makes sense to speak of the technology of the telephone allowing people to have long-distance relationships. It also makes sense to discuss the technical aspects of telephones in comparison to the postal system. Both the phone and the mail as technology mediate human relationships in the same way insofar as they allow us to communicate at great distances. Yet as technical phenomena they are completely distinct. To pose questions with respect to technology is to pose cultural and relational questions. To pose questions with respect to the technical is to pose instrumental questions. Put differently, to answer the question, What is hacking? properly, we cannot simply examine the manner in which hacking is done, the tools used, or the strategies that hackers deploy — the instrumental forces that constitute hacking. Instead we must look at the cultural and relational forces that define the context in which hacking takes place.

Hackers and Hacking

Not long ago, being called a hacker meant only that one belonged to a group of technology-obsessed college or graduate students who worked tirelessly on the dual diversions of finding interesting ways around problems (often in ways that resembled Rube Goldberg machines) and perpetuating clever, but harmless, pranks. This "class" of technophile is characterized by a kind of "moral code," as documented by Steven Levy in his 1984 book, *Hackers*. The code, as Levy describes it, was "a philosophy, an ethic, and a dream," and it was constituted by six basic theses:

1. Access to computers — and anything which might teach you something about the way the world works — should be unlimited and total. Always yield to the Hands-On Imperative!

2. All information should be free.

3. Mistrust Authority — Promote Decentralization.

4. Hackers should be judged by their hacking, not bogus criteria such as degrees, age, race, or position.

5. You can create art and beauty on a computer.

6. Computers can change your life for the better.[8]

The hackers Levy refers to were the original champions of the information superhighway, and their ethic was utopian in nature. As Levy describes it: "To a hacker a closed door is an insult, and a locked door is an outrage. Just as information should be clearly and elegantly transported within the computer, and just as software should be freely disseminated, hackers believed people should be allowed access to files or tools which might promote the hacker quest to find out and improve the way the world works."[9]

The "old hacker" of the 1960s and 1970s is often characterized with no small amount of nostalgia and is frequently seen as a counterpoint to the emergence of the new breed of hacker, the "cyberpunk" or "cracker." The "old hackers," in this romanticized telling, were "a certain breed of programmers who launched the 'computer revolution,' but just can't seem to be found around anymore. . . . [A]ccording to these 'old-school' hackers, hacking meant a willingness to make technology accessible and open, a certain 'love affair' with the computer which meant they would 'rather code than sleep.' It meant a desire to create beauty with computers, to liberate information, to decentralize access to communication."[10] In short, the old-school hacker was dedicated to removing the threat of high technology from the world by making that technology accessible, open, free, and "beautiful." To the 1960s hacker, hacking meant rendering technology benign, and hackers themselves not only were considered harmless but were framed as guardians of technology — scientists with an ethic that resembled Isaac Asimov's "Laws of Robotics": above all else, technology may never be used to harm human beings. Moreover, these hackers effected a strange anthropomorphism — information began to be personified, given a sense of Being usually reserved for life-forms. The old-school hacker was frequently motivated by the motto "Information wants to be free," a credo that attributed both a will and an awareness to the information itself. Interestingly, it is these two things, will and awareness, that seem to be most threatened by the evolution of technology. In an era when the public is concerned both with a loss of freedom to technology and with a fear of consistently finding themselves out of touch with the latest technological developments, there is a transference of our greatest fears about technology onto the idea of information. The hacker ethic remedies these concerns through the liberation of information. The logic is this: if technology cannot even confine information, how will it ever be able to confine us? Within the framework of this initial question we can begin to trace out the history of hacking as a history of technology.

A Genealogy of Secrecy

One of the primary issues that hackers and hacker culture negotiates is the concept of secrecy that has evolved significantly and rapidly since World War II. Indeed, hackers' relationships to technology can be understood as a cultural phenomenon and cultural response to the evolution of secrecy, particularly in relation to the broader political and social climate, the birth, growth, and institutionalization of the computer industry, and the increasing import of multinationalism in industry. The concept of secrecy seems to change from generation to generation. What secrecy means and particularly its value shift as social, political, and economic contexts change over time, but what has always remained stable within hacker culture is the need to negotiate the relationship between the technical aspects of the machines themselves and the cultural value of secrecy.

One of the first connections between secrecy and machines arose during the Allies' work to break German codes during World War II. Until this point, most cryptography had utilized methods of simple substitution, meaning that letters in the alphabet would be substituted for other letters, scrambling a clear text message into a ciphertext. For example, substituting the letter *a* for the letter *r, b* for *e,* and *c* for *d* would produce the ciphertext "abc" for the word "red." The problem with such a system of simple substitution is that the English language tends to utilize letters with fairly regular frequencies, and for that reason, no matter what substitutions are made, it becomes fairly easy to guess what has been encoded just by knowing how often certain letters appear in the message. Machines that encoded or decoded substitution schemes only helped speed up the process of encoding and decoding; they didn't actually perform the act of encoding in a meaningful way. That would all change with the German Enigma Machine, the first machine that actively encoded messages. The Enigma Machine consisted of eight code wheels that would rotate each time they were used. That meant that each time a substitution was made, the wheel making that substitution would rotate forward so the next time that letter was to be used, the substitution would be different. To complicate things further, each of the eight wheels had a different set of substitutions. The only way to decode the message was to have an Enigma Machine on the other end, with the wheels set to the original position, and feed the message back through to decode it.

The process of World War II code-breaking spawned the first generation of computer scientists and committed the evolution of the machine to the interests of secrecy and national security. From World War II on, two of the primary functions of the computer would be code-making and code-breaking. Indeed, it is not too far a stretch to claim that the first computer scientists (including Alan Turing, the "father" of modern computer science) who broke the Enigma Machine's coding process were, in the most basic sense, the first computer hackers and that the origin of computers rests with the need to keep and, perhaps more important, break secrets. The breaking of the Enigma codes would lead to the development of Colossus, what the museum at Bletchley Park (the site in Britain where the Enigma cipher was broken) calls "the world's first computer."

Secrecy has always been a significant concern for the military. Ever since Caesar scrambled messages sent to his troops, the military has recognized the need for secrecy. In the late 1930s and early 1940s, the stakes of military secrets were raised enormously with the invention and, ultimately, the use of the first atomic weapons. In that climate, the Department of Defense in the late 1950s turned to universities as the means to advance the study of computer science and engineering and, in turn, spawned what is generally acknowledged as the first generation of computer hackers. Funded almost exclusively by the Department of Defense, hacking began its difficult and oftentimes contradictory relationship to secrecy. As opposed to their forerunners, who worked exclusively in secrecy as they broke codes at Bletchley Park, hackers of the 1950s and 1960s, who worked in an environment of learning and academic freedom on university campuses, abhorred the notion of secrecy. And it was, in large part, that distaste for secrecy that led to most of the major advances that hackers would later make in the computer labs of MIT and other universities. "A free exchange of information," Levy writes, "particularly when the information was in the form of a computer program, allowed for greater overall creativity."[11]

Such an ethic led to cooperation among programmers and a nearly constant evolution of ideas and programming techniques. The goal was not just to write original programs but to improve on the work of others. The most skillful hack was not writing a new line of code but finding a way to do something in someone else's code that no one had seen before — to eliminate code, making the program run faster and more elegantly. While the ethic belonged to the hackers, the product belonged to the Department of Defense. That conflict, which most hackers of the 1950s and 1960s were either sheltered from or in denial about, represented the ultimate irony — work produced in the climate of absolute freedom would be deployed by the military in absolute secrecy. One hacker described the reaction to the realization that the funding and the results of those funded projects were military: "I got so upset I started crying....Because these people had stolen my profession. They had made it impossible to be a computer person. They sold out. They sold out to the military uses, the evil uses, of the technology. They were a wholly owned subsidiary of the Department of Defense."[12] Although these hackers were the "heroes of the computer revolution," as Levy argues, they were also the brain trust that developed most of the hardware and software that would drive the military-industrial complex throughout the late twentieth century and create innovations that would be deployed in the name of secrecy and surveillance.[13] The climate that hackers envisioned as "the realization of the hacker dream with sophisticated machines, shielded from the bureaucratic lunacy of the outside world," was only realized up to a point. All the late-night hacks, high jinks, and clever pranks that characterized the 1960s hackers were furthering another agenda, which would have wider cultural impact than the hackers working on those projects imagined. An example that Levy cites is the funding of speech recognition by the ARPA (the Department of Defense's Advanced Research Project Agency), a project that "directly increased the government's ability to mass-monitor phone conversations at home and abroad."[14] Perhaps most significant was ARPA's funding of ARPAnet, the foundation of the modern-day Internet, which was originally designed as a decentralized communication network intended to maintain command, control, and communication abilities in the event of nuclear war. In essence, these early hackers designed a secure communications system funded by the Department of Defense to ensure survivability in nuclear war.

To the hackers of the 1960s, secrecy meant the freedom to share code in the computer lab, the spirit of cooperation in program design, and the right to tinker with anything and everything based on one's ability to improve upon it. For instance, "In a perfect hacker world," Levy writes, "anyone pissed off enough to open up a control box near a traffic light and take it apart to make it work better should be perfectly welcome to make the attempt."[15] Freedom and secrecy were decontextualized to the point of solipsism. The hacker ethic, a matter of secrecy and freedom, was confined to the labs where hackers spent most of their waking hours and was naive about or ignorant of the greater context that was allowing that ethic to flourish. The shift that would make that naïveté most apparent was the move from the university to the corporate world. The second generation of hackers, the hackers of the 1980s, viewed secrecy differently. Without military funding and without corporate secrets to protect, these hackers took up the spirit of the original hacker ethic, but from a point of view that fully contextualized it in terms of politics, economics, and cultural attitudes. These hackers held to the tenets of freedom and abhorred the notion of secrecy, just as their predecessors did, but they lacked the solipsistic environment of the 1960s computer lab or the nearly unlimited government funding that made the original hacker ethic not only possible but also relatively risk-free.

Yesterday's Hackers

These original hackers of the 1950s and 1960s are generally recognized as the ancestors of the modern computer underground. There was, however, a second strain of hacker, one that is much more closely allied with the tradition of contemporary hacking. As Bruce Sterling writes, "the genuine roots of the modern hacker underground probably can be traced most successfully to a now much-obscured hippie anarchist movement known as the Yippies."[16] The modern underground, then, had its roots in a leftist political agenda that grew out of 1960s counterculture and was fueled by the antiwar protest movements of the 1960s and 1970s. "The Yippies, who took their name from the largely fictional 'Youth International Party' [YIP], carried out a loud and lively policy of surrealistic subversion and outrageous political mischief. Their basic tenets were flagrant sexual promiscuity, open and copious drug use, the political overthrow of any powermonger over thirty years of age, and an immediate end to the war in Vietnam, by any means necessary."[17] Jerry Rubin and Abbie Hoffman were two of the most high-profile members of YIP, but it would be two hackers, "Al Bell" and "Tom Edison," who would take over the newsletter (initially called *Party Line*) and transform it into *TAP* (*Technical Assistance Program*). The vision of *TAP* became increasingly less political as it began to focus increasingly on the technical aspects of telephony. Ultimately, *TAP*'s primary mission became the distribution of information, for example, "tips on such topics as lock picking, the manipulation of vending machines, do-it-yourself payphone slugs and free electricity."[18] Although the war in Vietnam had served as the origin for *TAP*, by the time the war had ended, *TAP* had become a technical journal completely divorced from politics.

In the late 1960s and early 1970s, while the hackers of *TAP* were busily disseminating information, hackers at MIT were busy creating information. The majority of funding for most, if not all, of these hackers' projects was coming from ARPA. The entire purpose of ARPA, begun in the early 1960s, was to create military applications for computers and computer networks. Although most of the hackers working at MIT either were blind to such considerations or found them irrelevant, hackers of the 1960s at major research universities were a significant portion of the technological side of the military-industrial complex (MIC). Hacker involvement in government projects with military application did not escape the attention of students protesting the war in Vietnam, and as a result computer labs at some major universities were the site of major protests and wound up being shielded by steel plating and half-inch-thick bulletproof Plexiglas.[19] The role of the MIC was central, although unintentional and unanticipated, to the formation of what constitutes the computer underground. On the one hand, the MIC produced an underground coming out of the protest movement of the 1960s. On the other hand, the MIC was funding the projects that would create hacker culture. Partially out of naïveté and partially because of the novelty of computers, the hackers of the 1960s and 1970s were able to avoid the obvious contradiction between their highly antiauthoritarian mind-set ("Information wants to be free") and the fact that the people they were designing systems and software for were not likely to respect that basic tenet. Because they were producing the vast majority of new technology, the old-school hackers were able to maintain the illusion that they were also controlling it. Within a decade, the "old school" had moved to the Silicon Valley and started to build an industry that would look and operate increasingly less like the labs at MIT and Harvard and more like the corporations and organizations against which the 1960s hackers had rebelled.

Hacking the Industry:
From the Altair to the Apple

Bill Gates: Cocky wizard, Harvard dropout who wrote Altair BASIC, and complained when hackers copied it.

Steven Jobs: Visionary, beaded, non-hacking youngster who took Wozniak's Apple II, made lots of deals, and formed a company that would make a billion dollars.

Steven "Woz" Wozniak: Openhearted, technologically daring hardware hacker from San Jose suburbs, Woz built the Apple Computer for the pleasure of himself and friends.
— Steven Levy, "Who's Who," from *Hackers*

It may seem odd to think about hacking and hacker culture in relation to three of the most important figures from the personal computer (PC) industry. The computer industry has always been and in many ways continues to be the very antithesis of hacker culture, which is also the reason that it plays an important role in the formation of hacker culture. Without the hackers of the 1960s, there never would have been a PC; without the PC, there never would have been a PC industry; without a PC industry, there never would have been the hackers of the 1980s and 1990s. Without a military-industrial complex, which funded most of the computer research in the 1950s, 1960s, and 1970s, it is likely there would never have been a PC or a hacker culture. One cannot hope to understand the hacker in his or her modern incarnation without understanding at least a little bit about where the PC originated.

Steven Jobs and Steven Wozniak, inventors of the first mass-marketed PC, get a lot of mileage out of the story that they began as hackers, or, more specifically, as phone phreaks. Phone phreaks are to telephones what hackers are to computers — they possess a basic understanding of how the phone system works and as a result can do things such as place long-distance calls for free. One of the first phone phreaks was John Draper, aka Captain Crunch, who took his handle from his discovery that in the early 1970s the whistle that came with Capt'n Crunch cereal sounded the tone (2600 Hz) that would allow one to take control of the phone line and place long-distance calls for free. Not long after that, this discovery was harnessed in a more technical manner through what became known as a "blue box," a small electronic device that would emit the 2600 Hz tone.

In contrast to Draper, Jobs and Wozniak learned about blue-boxing from a 1971 *Esquire* article[20] and built their own box. They not only used the boxes to make free calls but also went on to sell them to students in the Berkeley dorms.[21] In short, Jobs and Wozniak copied a device someone else had originated and then *sold* those copies. This is a pattern Jobs and Wozniak would follow in building the first Apple. In contrast to the long-standing hacker ethic that freely distributed information and knowledge and resisted the impulse toward commodification, Jobs and Wozniak openly embraced it.

Before there was the Apple, however, there was the Altair. The Altair was the very first PC. It came in parts, cost between four hundred and five hundred dollars, and had to be built (and not just assembled, but soldered) by the end-user. These limitations meant that PCs were limited to computer hobbyists, and most of those hobbyists belonged to the first generation of old-school hackers. These were computer enthusiasts who did everything from soldering the wires to programming the machine. One of those enthusiasts was Bill Gates. Gates, along with his friend Paul Allen, began writing the first language for the Altair, what would become Altair BASIC. BASIC, however, was not Gates and Allen's invention, but rather a programming language that had been put into the public domain a decade earlier by Dartmouth researchers Thomas Kurtz and John Kemeny. Gates and Allen's contribution was making BASIC, a language designed for a large mainframe computer, run on the Altair.

Almost all of the software for the Altair was written by hobbyists, who routinely shared their programs with other hackers at meetings of computer clubs across the country. The sharing of information became one of the central tenets of the hacker ethic, and this would be the central organizing principle of one of the first computer clubs, the Homebrew Computer Club of Menlo Park, California. Later, as the early pioneers of hacking would make their mark in the corporate world, a new ethic would emerge that would change everything. The transformation was intensified as the process of commodification increased. With commodification, the earliest computer hackers, those who had built their computers themselves and shared every tidbit of information that might help to improve their machines or programs, were in competition with each other, fighting to create and to maintain market share. The result was dramatic — competition in the marketplace "retarded Homebrew's time-honored practice of sharing all techniques, refusing to recognize secrets, and keeping information going at an unencumbered flow.... All of a sudden, they had secrets to keep."[22] That transition marks the dividing point between the old-school hackers of the 1960s and 1970s and the new-school hackers of the 1980s and 1990s.

The Hacker Imagination: From Sci-Fi to Cyberpunk

It should come as no surprise that the hacker tradition is grounded in the literature of science fiction and fantasy. Internet culture has its roots in the first e-mail discussion list that emerged early in the life of ARPAnet (the precursor of today's Internet). That list was SF-LOVERS, a list of people devoted to the discussion of science fiction. But like the shift from the 1960s hackers to the 1990s hackers, sci-fi literature also underwent a radical shift. The 1960s hacker's inspiration was found in the literature of Isaac Asimov, Philip K. Dick, Norman Spinrad, or Harlan Ellison, all writers who depicted a future of possibility, who wrote cautionary tales. These tales usually begin in an unfamiliar world, one that isn't your own but threatens to be. Often set in the future, but not too far in the future, these novels present anything from an alternative history to a fantasy world where the strange and unusual are commonplace. For example, Philip K. Dick's 1969 novel *Ubik* begins with the line, "At three-thirty a.m. on the night of June 5, 1992, the top telepath in the Sol System fell off the map in the offices of Runciter Associates in New York City."[23] Dick's world is both familiar and strange — it contains elements we know and some we do not know. New York and all the familiar conventions are intact — time, date, and a business with a realistic name. Yet the description also points to the unfamiliar: in this new world telepathy is not only possible but has become a commodity. We are, in essence, introduced to a world of possibility that is familiar enough to be recognizable yet strange enough for us to take notice of the ways in which our future might change.

In comparison, the mainstay of the 1990s hacker was the literature of cyberpunk, represented by William Gibson and Jon Brunner. Their novels are predominantly dystopic, describing a battle that has already been fought and lost. Consider Gibson's opening to *Neuromancer:* "The sky above the port was the color of television, tuned to a dead channel."[24] Immediately we know that Gibson's vision is going to challenge the basic model of the cautionary tale. From the outset, what it presents is threatening, and the dystopia of Gibson's fiction is taken as preordained. The literature of cyberpunk so dominated the imagination of the 1990s hackers that, in many ways, they came to see themselves as antiheroes, based on the prototype of Gibson's characters and others. These characters live in a world defined, even in its geography, by information. In *Neuro-*

mancer, Gibson describes a world that the hackers of today have adopted as their own:

> Home.
> Home was BAMA, the Sprawl, the Boston-Atlanta Metropolitan Axis.
> Program a map to display frequency of data exchange, every thousand megabytes a single pixel on a very large screen. Manhattan and Atlanta burn solid white. Then they start to pulse, the rate of traffic threatening to overload your simulation. Your map is about to go nova. Cool it down. Up your scale. Each pixel a million megabytes. At a hundred million megabytes per second, you begin to make out certain blocks in midtown Manhattan, outlines of hundred-year-old industrial parks ringing the old core of Atlanta.[25]

In such an account, we can begin to see the manner in which the exchange of information and data begins to code more common points of reference. Not only have old, familiar locales (Manhattan, Atlanta) been recoded as data, but our entire way of seeing them has been transformed. The way we look at a map no longer is about distance but rather is about the density of information. The map not only is not the territory — it has lost its relationship to the territory in terms of representation. The map is no longer a scale model of space. What is represented is the "frequency of data exchange." This is a revealing term. "Frequency," a term of temporality, not location, is the first focus, and "exchange" is the second. What is lost in this "mapping" is any sense of place. Place is erased, making it possible for Gibson to write of "home" as a place without place, home as a system of exchange.[26]

The literature of cyberpunk teaches, or reflects, the value of information both as data and as a social fabric, a medium of exchange, and a relational concept. Cyberpunk represents a world where information has taken over, and the literature provides a sense of the fears, dangers, anxieties, and hopes about that new world. As opposed to the earlier hackers who sought to "liberate" information, hackers of the 1990s see themselves as trapped by information. As a result, the "hacker ethic," which took as its most basic tenet that "information wants to be free," needs to be radically transformed. While the spirit of the ethic remains intact, the letter of it must necessarily change. Where the hackers of the 1960s had a great deal of control over the information that they created and utilized, the hackers of the 1990s and the beginning years of the twenty-first century find themselves in a world so overwhelmed by information that control itself becomes the contested issue. Most important to the later hackers is the concept that information is now their home, and secrecy of information is the equivalent of confinement or prison. The original ethic is, for the most part, still intact, but its meaning, value, and application have been radically altered by the ways in which the world has changed.

The cyberpunk vision of the future has radically reshaped the vision of the latter-day hacker. In one document currently being circulated on the Internet, the "Declaration of Digital Independence," one hacker describes the Internet as the next battleground for the regulation of information, and hence freedom. "It [the Internet]," he writes, "should be allowed to make its own rules. It is bigger than any world you can and can't imagine, and it will not be controlled. It is the embodiment of all that is free; free information, friendship, alliances, materials, ideas, suggestions, news, and more."[27] The hackers of the 1960s, inspired by the utopian science fiction of their day, saw the battle in terms of free information and felt encouraged by that literature to experiment, learn, and develop. In contrast, the hackers of today, with the dystopic vision of cyberpunk, see this battle as already lost and as something that can only be rectified by revolution. In part, this is the result of the increasing commodification of information, which has created a media that they describe as "the propaganda vending machine of today," which, "as a whole, trip over themselves, feeding lies to the ignorant."[28]

These hackers hold that the commodification of information has led to an increasing investment of power in the media. Accordingly, they argue, a transformation has taken place. The media, who "enjoy the power of managing information" in an era of commodification of information, are no longer interested in the freedom of information, but, rather, are invested in the careful control and dissemination of information. As Paul Virilio argues, the commodification of information has had an ironic effect: "[T]he industrial media have gone the way of all mass production in recent years, from the necessary to the superfluous. . . . [T]heir power to denounce, to reveal, to flaunt has been growing endlessly to the detriment of the now precarious privilege of dissimulation — so much so that currently the real problem of the press and television no longer lies in what they are able to show as much as in what they can still manage to obliterate, to hide."[29] The point is not so much that the media do not break news or reveal secrets as much as it is that they are selective about which secrets can and will be revealed. As media power is increasingly consolidated, media outlets have a possessive investment in particular stories and a similar investment in keeping other stories quiet. The more centralized the media become, the more power they have to self-regulate what constitutes "news."

Today's hackers, who follow this dystopic vision, contrast the media's approach to the management of information with their own sense of boundless curiosity. As the "Digital Declaration of Independence" illustrates: "Everyone has the need to know, the curiosity of the caveman who invented fire, but some have been trained like monkeys, not ever knowing it's there. They simply accept things, and do what is expected of them, and this is sad. They are those who never fight back, and never open their minds. And they are, unfortunately, usually the governing bodies; the teachers, bosses, police, federal agents, congressmen, senators, parents, and more. And this, my friend, must change."[30] Such a call to action defines the problem in terms of curiosity and recognizes that such curiosity is problematic within contemporary culture. Curiosity becomes dangerous and even subversive not to any particular group or organization but in principle. Curiosity is precisely what threatens secrecy, and in doing so, it challenges the economic structure, the commodification, of information.

Blurring the Lines between Old and New: *WarGames*, Robert Morris, and the "Internet Worm"

In 1988, the distinction between the old and the new hacker was clearly staked out. This difference was revealed through two figures: David Lightman, the protagonist from the film *WarGames* and the prototype for the new-school hacker; and Robert Morris, a quintessentially old-school hacker.

The new-school hacker was introduced into the popular imagination through the 1983 release of the film *WarGames*, featuring Lightman (played by Matthew Broderick) as a curious kid exploring computers and computer networks who unwittingly starts the U.S. military on the road to World War III by playing what he thinks is a game — global thermonuclear war.

WarGames opens, somewhat ominously, with a scene in which U.S. soldiers in missile silos are ordered to fire their weapons at the Soviet Union, beginning what they believe will be the third, and undoubtedly last, world war. The soldiers are uncertain as to whether the order is part of a training exercise or not. As a result, a large percentage of the soldiers, uncertain about the effects of their actions, choose not to fire their missiles. The orders are part of a simulation, designed to test U.S. military battle-readiness. Their failure results in the implementation of a new program, wherein humans are removed from missile silos and replaced by electronic relay switches and strategic decisions about nuclear warfare are to be made by a state-of-the-art computer named WOPR (pronounced whopper) — War Operations Planned Response. The machine is devoted to constantly replaying World War III in an effort to maximize the effectiveness of U.S. missiles and minimize U.S. casualties. In effect, its job is to figure out how to win a nuclear war.

The young hacker, David Lightman, stumbles across WOPR quite accidentally while searching for a game company called Protovision. Using a modem and computer program (since termed a WarGames dialer), the hacker scans every open phone line in Sunnyvale, California (in the heart of the Silicon Valley), looking for modem access to Protovision. He comes across a system and, through hours of research (principally learning about the system's designer, Stephen Falken) and hacking, gains access.

The initial game continues to run even after Lightman is disconnected from the system, and, as Lightman soon discovers, the computer is unable to distinguish the game from reality. In less than three days time, WOPR will calculate a winning strategy and fire its weapons. In the meantime, Lightman is arrested. On the verge of being charged with espionage, he escapes custody and searches out the system designer with the help of his girlfriend, Jennifer Mack (played by Ally Sheedy), to convince him to help persuade NORAD (North American Air Defense command) not to believe the information that the computer is sending them. Falken, the system designer, has become a recluse since the death of his son, Joshua (whose name is the secret password that gains Lightman access and the name that Lightman and Falken use to refer to the WOPR computer). Eventually, after several incidents that put the world on the brink of a nuclear holocaust, Lightman manages to teach the computer (now Joshua) the meaning of futility by having it play ticktacktoe in an infinite loop sequence. The computer concludes, after exhausting the game of ticktacktoe and learning from it, that nuclear war is also unwinnable and that "the only winning move is not to play."

The film demonstrates a tremendous anxiety about technology, represented both by the missiles that threaten to destroy the United States and the Soviet Union and by the machines that control those missiles.[31] The hacker, however, is represented in a more ambivalent manner. On the one hand, the harmlessness of Lightman's actions early on (using his computer to make faux airline reservations or even finding a bank's dial-in line) is made clear by his good nature and curiosity. That sense of "playing a game" is radically transformed when the machine he hooks up to (the WOPR) is unable to distinguish between a game and reality. On the other hand, Lightman and his understanding of "playing a game" (in this case ticktacktoe) ultimately are able to save the day.

In the film, the hacker is positioned as dangerous because he is exploring things about which he has little or no understanding. It is easy in a world of such great technical sophistication, the film argues, to set unintended and potentially disastrous effects into motion even accidentally. But equally important is the characterization of the hacker as hero. WOPR, above all else, is a thinking machine, an artificial intelligence, and that thinking machine needs guidance and instruction in its development. Technology is infantilized in the film (underscored by the use of the name of Falken's deceased son, Joshua), and the message of the film is that Lightman, the hacker, is the most appropriate educator for the technology of the future. Not the generals, the system administrator, or even the system designer himself is able to teach the machine, but Lightman, the hacker, can. The hacker stands at the nexus between the danger and the promise of the future of technology.

Thus, the intersection of hacker culture and popular culture is clearly and conspicuously marked. With the release of *WarGames,* hacker culture had a national audience. That culture, however, was as much a product of the film (and the response to the film) as it was a reality. While there certainly had been a long history of hacking and phreaking that predated *WarGames,* the hacking community itself was small, exclusive, and rather inconspicuous. With *WarGames* that all changed. As Bruce Sterling describes it, "with the 1983 release of the hacker-thriller movie *WarGames,* the scene exploded. It seemed that every kid in America had demanded and gotten a modem for Christmas. Most of these dabbler wannabes put their modems in the attic after a few weeks, and most of the remainder minded their Ps and Qs and stayed well out of hot water. But some stubborn and talented diehards had this hacker kid in *WarGames* figured for a happening dude. They simply could not rest until they had contacted the underground — or, failing that, created their own."[32] To varying degrees, hackers themselves admit to this, although none would probably state it in precisely Sterling's terms. One of the primary differences between the film's depictions and Sterling's is that the film *WarGames* has absolutely no representation, or even suggestion of, an underground.

Undoubtedly, the film had a greater impact on hacker culture than any other single media representation. Hackers such as Shooting Shark and Erik Bloodaxe, in discussing their early influences, both confess (somewhat reluctantly) that the film had a major impact on them. "Embarrassing as it is for Erik," his Pro-Phile reads, "*WarGames* really did play a part in imbedding the idea of computer hacking in his little head. (As it did for hundreds of others who are too insecure to admit it.)"[33] Shooting Shark's chagrin is equally obvious, "Worse yet, '*WarGames*' came out around this time. I'll admit it, my interest in hacking was largely influenced by that film."[34] Many hackers took their handles from the film, including "David

Lightman" and a whole spate of Professor Falkens or, more often, Phalkens, named for the computer genius who invented the WOPR. The film has even been held as the inspiration for one of the most serious instances of hacking in the 1980s, the exploits of a West German hacker who infiltrated U.S. military computers and sold U.S. government information to the Soviets.[35] The fact of the matter is that while many teens may have been intrigued by the possibility of breaking into high-security military installations, that aspect was never glamorized in the film. More likely, what intrigued young hackers-to-be was an earlier scene, where Matthew Broderick logs on to the local school's computer using a modem from his PC at home and changes his grades in order to avoid having to take classes in summer school. The lure for the young hacker was never with starting World War III, but rested, instead, with the ability to make local conditions (for example, school work) more tolerable.

This narrative presents the good-natured, well-intentioned hacker innocently wreaking havoc as a result of his explorations. If Lightman is the introduction of the hacker to the popular imagination, Robert Morris would be the turning point in the perception of what kinds of threats hackers pose to society.

Perceptions of who and what hackers are underwent another transformation in the late 1980s, and that moment can be marked quite clearly in the popular imagination by a single case — the day the Internet shut down, November 8, 1988. Although the Internet was only a fraction of the size that it is today, it was an information infrastructure that had grown large enough that many government agencies (not the least of which was the military) and especially colleges and universities had come to rely on it. *WarGames* had provided the fantasy, the cautionary tale, but it all had ended well. There had been no disaster, and, even though the hacker had created the problem himself, it was also the hacker who saved the day, teaching the computer the lesson of futility by giving it a game (ticktacktoe) that it couldn't win, by becoming the educator of the technology of the future. The hacker, in the old-school tradition, saved the day by rendering technology itself (in that case WOPR) benign. Such was not the case with Robert Morris.

In 1988, Morris launched what has come to be known as

the "Internet worm," a computer program that transmitted itself throughout the Internet, eating up an increasing number of computing cycles as it continually reproduced itself. Morris's intent was to have the worm reproduce at a rate that would allow it to continue throughout the system, unnoticed for years. With each system it entered, it would exploit a security hole, discover user passwords, and mail those passwords back to Morris. Morris, however, made a small error that had a great effect, and the worm reproduced itself at a rate much faster than he had anticipated. Rather than telling the worm to stop running when it encountered a copy of itself, which would have stopped it from spreading, Morris coded the worm so that one in every seven worms would continue to run even if other copies of the worm were found on the machine. The ratio was too high "by a factor of a thousand or more," and Morris's worm replicated itself at an astonishing rate, essentially eating up computing cycles at such a rate that the machines could do nothing but run Morris's program.[36]

Morris's worm spread throughout the nation's computer systems at an alarming rate. The day after the worm was released, a significant portion (no one is really sure of the exact number) of those computer systems hooked to the Internet had ground to a dead stop. The only remedy was to disconnect from the network and wait for the experts working around the country to find a way to counteract the program.

What makes this case special — apart from the fact that it focused national attention on issues of computer security, Net vulnerability, and the degree to which we have come to depend on computers in our daily lives — was the ambivalent stature of the programmer himself. By all rights, Robert Morris was an old-school hacker. A computer-science graduate student at Cornell, he had spent most of his college career (and early years) working with computers, devising clever workarounds for difficult problems, and engaging with his special fascination, computer security. The son of Bob Morris, the chief scientist at the National Security Agency's National Computer Security Center, Robert learned the ins and outs of computer and system security at a very early age and possessed a natural talent for finding system bugs and holes he could exploit to gain access.[37]

Even the worm itself was never intended as anything more than an exposition of the flaws in UNIX system design.

Morris's prank, however, was extraordinarily similar to the most malicious activity that was being attributed to the new breed of hackers, the computer virus. The difference between a worm and a virus was hotly debated within the computer community.[38] Worms, which are independent programs that move under their own power, are generally considered to be beneficial things. First developed by researchers at Xerox, worms were programs that would run throughout a system performing useful tasks. The name is taken from John Brunner's 1975 novel, *Shockwave Rider,* where the protagonist feeds a "tapeworm" into the government's computer, as a means to counter government surveillance. The act, in Brunner's case, is one of heroism in the face of an oppressive, tyrannical government. In that sense, worms carry with them the connotation of performing a useful function, usually in line with the traditional hacker ethic. Many were reluctant, it seems, to label Morris's program a worm because of the positive connotations that the term carried.

Viruses, in contrast, are generally considered to be malicious. The connotations of viruses as sickness, illness, and even death (particularly in the age of AIDS and Ebola) provide an interesting counterpoint to the discussion of the "Internet worm." Many who argued for the classification of Morris's program as a virus did so on the ground that it was harmful and caused great damage. The community was split both on how to refer to the program and on its ethical implications. The line between the old school and the new school had begun to dissolve.

Morris's worm program spurred a great deal of speculation about his intentions and influences (Morris refused to talk with the press) and a great deal of debate over the threat that technology posed to society if it fell into the wrong hands. As Katie Hafner and John Markoff put it, "it also engaged people who knew nothing about computers but who were worried about how this new technology could be used for criminal ends."[39] In the popular imagination, the line between the old and new hacker was already fuzzy, and Morris's worm program only made the distinction fuzzier.

A second drama was also being played out around anxieties

about technology. Those anxieties coalesced around the notions of secrecy and technology and were reflected in popular culture's representations of hackers.

For example, films about hackers almost always deal with the question of secrecy. If we are to take *WarGames* as the prototypical hacker film, it is easy to see exactly what this means. *WarGames* begins with David Lightman trying to break into Protovision in an effort to play the latest games before they are publicly released. In essence, Lightman seeks to break Protovision's corporate secret in order to have access to "secret" games. As the name suggests, Lightman wants to see something before anyone (or, more to the point, everyone) else does. While searching for Protovision's phone number, Lightman comes across a more interesting system, one that "doesn't identify itself." Thinking that this is Protovision's dial-in number, Lightman attempts to hack into the system, even after he is warned by two system programmers that the system is "definitely military."

The crucial point about the basic theme underwriting the film is that these two cultures of secrecy, Protovision's and NORAD's, are virtually identical. The structure of *WarGames* depends on our understanding and acceptance of the confusion of a corporate computer-game manufacturer's notion of secrecy with that of NORAD. In other words, corporate and military secrets are, at some level, indistinguishable. In the films that followed *WarGames,* this theme was renewed and expanded upon. In *Sneakers* (1992), the confusion between corporate and government secrecy is complete when it is revealed that the only real governmental use for the "black box" agents have been sent to recover is to snoop on other U.S. governmental departments, rather than on foreign governments. What we witness throughout the film, however, is that as a corporate tool, that black box is capable of everything from corporate espionage to domestic terrorism. Indeed, the project "SETEC ASTRONOMY" which is at the center of *Sneakers,* is an anagram for "TOO MANY SECRECTS." And while it is a government project (funded by the NSA in the film), it is revealed in the film's epilogue that the box will not work on other countries' cryptographic codes, only on those of the United States. Accordingly, the message is clear: domestically, there are to be "no more secrets" kept from those in power.

In two films of the 1990s, *Hackers* (1995) and *The Net* (1995), secrecy plays a major role as well. In these films, however, secrecy is what allows criminality to function in both the government and corporate worlds. In the case of *Hackers,* two employees (one a former hacker) of a major corporation are running a "secret worm program" to steal millions of dollars from the corporation. They are discovered when a hacker unwittingly copies one of their "garbage files" that contains the code for the worm's program. The same plot in played out in *The Net,* but from a governmental point of view. Angela Bassett (played by Sandra Bullock) accidentally accesses and copies secret governmental files that reveal wrongdoing on the part of governmental officials, again demonstrating the manner in which the culture of secrecy is able to hide and allow for a deeper sense of criminality to ferment and function. In both cases, hackers, by violating the institutions' secrecy, expose criminality by enacting criminality. The message from the later films is that secrecy creates a space for the worst kinds of criminality, which, because of the culture of secrecy, can only be exposed by another type of criminality — hacking.

Hackers of Today

As Steve Mizrach has noted, the split between the hackers of the 1960s and those of today is cultural and generational rather than technological:

The main reason for the difference between the 60s and 90s hackers is that the GenXers are a "post-punk" generation, hence the term, "cyberpunk." Their music has a little more edge and anger and a little less idealism. They've seen the death of rock n' roll, and watched Michael Bolton and Whitney Houston try and revive its corpse. Their world is a little more multicultural and complicated, and less black-and-white. And it is one in which, while computers can be used to create beauty, they are also being used to destroy freedom and autonomy.... [H]ence control over computers is an act of self-defense, not just power-hunger. Hacking, for some of the new

"hackers," is more than just a game, or means to get good-
ies without paying for them. As with the older generation, it
has become a way of life, a means of defining themselves as a
subculture.[40]

Born in the world that the 1960s hackers shaped, this new generation has been jaded precisely by the failure of the old-school
hackers to make good on their promises. Technology has not been
rendered benign, and information, while still anthropomorphized
and personified, reveals that we are more like it — confined, coded,
and organized — than information is like us. The cautionary tales,
like Asimov's, that guided the 1960s hackers have been replaced
with tales of a dystopian cyberpunk future that features technology no longer in the service of humankind, but humankind fused
with technology through cybernetics, implants, and technological
modifications to the body.

There seem to be two reasons for the shift. The first is that the rate
of technological growth has outstripped society's capacity to process
it. A certain technophobia has emerged that positions technology as
always ahead of us and that produces a fear that is embodied by the
youth of contemporary culture doing things with computers that an
older generation is unable to understand. Hacking promotes fear, but
it is about a contained kind of fear, one that is positioned as a form
of "juvenile delinquency" that these youth will, hopefully, grow out
of. In that sense, hackers emerge as a type of "vandal," a criminal
who is often malicious, who seeks to destroy things, yet is terribly
elusive. The threat, like the technology that embodies the threat, is
decentralized, ambiguous, and not terribly well understood, but it
doesn't need to be. We feel we can trust our information networks,
for the most part, the same way we can trust our trains and buses.
Occasionally, someone may spray paint them, flatten a tire, or set a
fire on the tracks, but these things are inconveniences, not disasters.
Hackers pose a similar type of threat: they may deface the surface
of things, but the underlying faith in the system remains intact. Just
as one does not need to understand how internal combustion en-
gines work to trust that a car will function properly, one does not
need to understand how information networks function in order to

use them. The second reason for a fear of hackers is the result of a displacement of anxiety that the hackers of the 1960s have identified — namely, the increasing centralization of and lack of access to communication and information. This new generation of hackers has come to represent the greatest fear about the 1960s dream of free and open information. John Perry Barlow, cofounder of the EFF (Electronic Frontier Foundation), wrote in the organization's manifesto, titled "Crime and Puzzlement," about his first interactions with two members of the New York hacking collective Masters of Deception. These two hackers, Phiber Optik and Acid Phreak, in an effort to demonstrate their online skills, threatened Barlow in one of their first encounters in a most unusual way:

> Optik had hacked the core of TRW, an institution which has made my business (and yours) their business, extracting from it an abbreviated (and incorrect) version of my personal financial life. With this came the implication that he and Acid could and would revise it to my disadvantage if I didn't back off.[41]

What is unusual about this threat is the manner in which it employs the vision of the 1960s hacker with a completely inverted effect. The threat is the removal of secrecy — a true freedom of information. It is important to note that TRW's report was not something that Barlow had ever consented to, nor was it something that he had any control over. In making it free, accessible, and open, the hackers posed the greatest threat — the ability to change the unchangeable, to access the secret, and to, in the process, disrupt a significant portion of one's life. If nothing else, the importance of the secrecy of information is documented (not coincidentally) by the hacker in the most dramatic fashion. Indeed, downloading TRW reports is a common trick that hackers will employ to startle, scare, and even intimidate media personalities, interviewers, and, in some cases, judges and lawyers. What Barlow identifies, with comic overtones, illustrates precisely what the implications of this reversal are. "To a middle-class American," Barlow writes, "one's credit rating has become nearly identical to his freedom."[42] That is, freedom relies on secrecy. The culture of information that the 1960s hacker feared has come to pass. In a kind of

Orwellian doublespeak, secrecy has become freedom, and the need for security (through implicit distrust of others) has become trust.

Old-school hackers, such Clifford Stoll, a Berkeley astronomer and system administrator, put forth the thesis that computer networks are and should be built on trust, a principle that seems in line with the 1960s hacker ethic. When confronted with a hacker who explained that he had hacked into Stoll's system to "show that your security isn't very good," the latter replied, "But I don't want to secure my computer. I trust other astronomers."[43] However, hackers of the 1990s argue that it is precisely this sort of argument that illustrates the hypocrisy of the 1960s hackers, many of whom have become rich in the computer revolution precisely by betraying the principles of openness, access, and freedom that they argued for as their ethic. The most interesting example is, perhaps, one of the most illustrative as well. As Steve Mizrach argues, "[Steven] Levy rants about those greatest hackers who founded Apple Computer and launched the PC revolution — those same ex-phreaks, Jobs and Wozniak, who actually allowed their company to patent their system hardware and software!"[44] To this day, Apple has been extremely successful in keeping both its hardware and software proprietary.

In essence, today's hackers argue, with a great deal of justification, that the hackers of the 1960s have become their own worst nightmare, the keepers of the secrets, those who block access to information and technology, all in the name of corporate self-interest. The new-school hacker, then, seems little more than the logical carryover from the earlier generation, a generation that spoke (and continues to speak) with such earnest commitment to the "good-old days" that they are unable to see how their own ethic implicates them in precisely that which they so fervently disavow. The connection, which the 1960s hackers are so loathe to make, is, in many ways, undeniable: "Indeed, the 90s hackers pay a lot of homage to the first generation. They have borrowed much of their jargon and certainly many of their ideas. Their modus operandi, the PC, would not be available to them were it not for the way the 1960s hackers challenged the IBM/corporate computer model and made personal computing a reality." In that sense, today's hackers are the children of the 1960s hackers, and that connection is not lost on the younger generation. In fact, today's hackers have inherited not only the tools from the older generation but much of their culture as well: "[T]heir style, their use of handles, their love for late-night junk food, are all testaments to the durability and transmission of 1960s hacker culture."[45]

They have also inherited, for the most part, their ethic. That ethic has been transformed, undoubtedly, but so have the conditions under which that ethic operates. These conditions are, in many ways, the progeny of the 1960s as well. Exploring, which seemed so harmless to the 1960s hacker, was only harmless because the culture of secrecy had not fully taken hold. As Barlow describes his in-person meeting with Phiber Optik, he "encountered an intelligent, civilized, and surprisingly principled kid of 18 who sounded, and continues to sound, as though there's little harm in him to man or data. His cracking impulses seemed purely exploratory, and I've begun to wonder if we wouldn't regard spelunkers as desperate criminals if AT&T owned all the caves."[46] This was the same hacker who had just days before threatened Barlow with a revision of his TRW credit report that threatened to destroy him financially.

How are we to explain this seemingly split personality? There are several reasons the current hacker comes across so brazenly, not the least of which is the proliferation of media in which hackers have begun to appear. While the 1960s hacker was confined to talking, predominantly, to other hackers, today's hacker is online in a world where there are few aspects of daily life that are not controlled or regulated by computers. Where computers were a novelty in the 1960s, today they are a desktop necessity. As computers entered the popular imagination, the hacker came along and was transformed with them.

A primary difference between the hackers of the 1960s and those of today rests with the fact that the latter are, for want of a better term, "media ready." For example, *CuD* (*Computer Underground Digest*), an online publication that tracks news of the underground, published an essay in response to a Fox News story on the "Hollywood Hacker" that eventually led to his arrest and prosecution. Although the essay does take up questions of the specifics of how computer raids are conducted, how warrants are obtained, and so on, the first two issues of concern, listed in the section "Why Should the CU Care?" are as follows: "1. The role of the media in inflaming public conceptions of hacking seems, in this case, to exceed even the cynical view of sensationalistic vested interest." Of equal concern is the "hacker hyperbole" that accompanied the Fox news report. "2. A second issue of relevance for the CU is the definition of 'hacker.'" What is at stake for the computer underground is the very control of the term "hacker" and what constitutes "hacking."[47]

To say that there is an awareness among hackers of how they are portrayed in the media would be a drastic understatement. Speaking of Phiber Optik and Acid Phreak, Barlow describes the phenomenon: "They looked about as dangerous as ducks. But, as *Harper's* and the rest of the media have discovered to their delight, the boys had developed distinctly showier personae for their rambles through the howling wilderness of Cyberspace." These personae are not merely inventions of the media, but are formed in a kind of cooperative venture between the media and the hackers themselves. "Glittering with spikes of binary chrome," Barlow writes, "they strode past the klieg lights and into the digital distance. There they would be outlaws. It was only a matter of time before they started to believe themselves as bad as they sounded. And no time at all before everyone else did."[48]

Books, articles, newspaper reports, films, and TV documentaries have tracked, with varying degrees of accuracy, the exploits of some of today's more high-profile hackers. One hacker, Kevin Mitnick, who has led authorities on several nationwide manhunts as a result of his hacking exploits, has been the subject of three books that detail his crimes and exploits. Others, such as Robert Morris, Phiber Optik, Kevin Poulsen, the members of the Legion of Doom, and the Masters of Deception have all been featured in books that range from the journalistic to high drama.

These new hackers have captured the spotlight in a way that the hackers of the 1960s never did. Old hackers captured attention in complete anonymity, for example, by sneaking into the Rose Bowl prior to the game and substituting the cards held by fans during the game so that they read "Cal Tech" rather than "Washington" (as they were supposed to). It made no difference to the hackers that Cal Tech wasn't even playing that weekend.[49] In contrast, in 1988, when AT&T suffered a major failure in long-distance telephone service that was found to be the result of a software glitch, reports immediately circulated that the interruption of service was the result of a hacker in the New York area who had broken into AT&T's computer system as a protest against the arrest of Robert Morris.[50] The media, as well as the public, have learned to expect the worst from hackers, and as a result, hackers usually offer that image in return, even if their own exploits are no more than harmless pranks or explorations grounded in curiosity.

If hacking is about imagination, then the reasons hackers hack are probably as numerous as the hackers themselves, and the means by which they accomplish their tasks range from the highest end of the technological spectrum to the lowest. But if we are to understand what hacking is and who hackers are, we need to separate out the people from the machines. To divorce hacking from its technical aspects, however, is not to divorce it from technology altogether. Hacking is about technology; arguably it is about nothing but technology. Hackers and hacking constitute a culture in which the main concern is technology itself and society's relationship to the concept of technology. Accordingly, I adopt the term "culture of technology" as a way to understand the cultural implications of technology from the hackers' point of view as well as from a broader cultural standpoint.

Hackers themselves rarely, if ever, talk about the tools they use. Indeed, their activity demonstrates that computer tools are above all mere vehicles for activity. In many cases, today's hackers utilize common UNIX servers as their goals, targets, and mechanisms for hacking — that is, they utilize other people's resources as a means to accomplish their goals. Even the most high-powered PC is little more than a dumb terminal that allows the hacker to connect with a more powerful corporate or university machine, an act that can be accomplished just as effectively with a fifteen-year-old PC or VT100 terminal as it can with the fastest, highest-end multimedia machines on the market today.

Hacking is not, and has never been, about machines, tools, programs, or computers, although all of those things may appear as tools of the trade. Hacking is about culture in two senses. First, there is a set of codes, norms, values, and attitudes that constitute a culture in which hackers feel at home, and, second, the target of hackers' activity is not machines, people, or resources but the relationships among those things. In short, hacking culture is, literally, about *hacking* culture. As culture has become dependent on certain types of technology (computers and information-management technology, particularly), information has become increasingly commodified. And commodification, as was the case with the first personal computer, is the first step in the revaluing of information in terms of secrecy. As Paul Virilio maintains, along with commodification comes a new way of valuing information — in "the maelstrom of information in which everything changes, is exchanged, opens up, collapses, fades away, gets buried, gets resurrected, flourishes, and finally evaporates in the course of a day," duration no longer serves as an adequate means of valuation. Instead, he argues, "speed guarantees the secret and thus the value of all information." Accordingly, American, and perhaps all of Western, culture's relationship to information has been undergoing radical change, moving from a culture that values duration to one that values secrecy. That transformation also marked the moment of emergence of the hacker, a moment Virilio situates as a "data coup d'etat" that originated with the "first military decoders to become operational during the Second World War." It was those machines, the "ancestors of our computers and software systems," that produced the merging of "information and data processing" with the "secret of speed."[51] And, indeed, Virilio is right in the sense that World War II produced the first machines that were capable of what we currently think of as cryptography and that were, perhaps, the beginning of the union of information and speed to produce secrecy. The employment of machines that were, in essence, rudimentary computers made complex coding and decoding efficient. This allowed for the production of codes that were much more complex and, therefore, much more difficult to break. As speed and information merge, secrecy becomes an increasingly important component of the culture in which we live. But such secrecy is precisely what hacker culture abhors.

Secrecy is not limited to encryption schemes but begins, as Virilio points out, with the process of commodification. In 1975, when hobbyists were busily programming the Altair, coding was done on paper tape and had to fit in 4K of memory. As noted earlier, one of the first successes was Altair BASIC, a programming language written by two college students, Paul Allen and Bill Gates, that would allow others to develop software for the Altair. The difference between Allen and Gates's Altair BASIC and just about every other program written for the Altair was that Allen and Gates sold their program, rather than giving it away. This difference was enormous, since for computer hobbyists, the question was never one of profit, but one of access. Dan Sokol, the person who had obtained and copied the original version of Allen and Gates's BASIC, distributed the program at the next meeting of the Homebrew Computer Club and "charged what in hacker terms was the proper price for software: nothing. The only stipulation was that if you took a tape, you should make copies and come to the next meeting with two tapes. And give them away."[52] In no time, everyone had a copy of Allen and Gates's program. Bill Gates responded by sending an "Open Letter to Hobbyists," which was published both in the Altair users' newsletter and in the Homebrew Computer Club newsletter. "As a majority of hobbyists must be aware, most of you steal your software," Gates wrote, accusing hobbyists of being "thieves." "What hobbyist," Gates continued, "can put 3 man-years into programming, finding all the bugs, documenting his product and distributing for free?"[53]

In essence, Allen and Gates treated the BASIC interpreter as a secret that could be purchased. Most other hackers didn't see it that way. And for them, ownership was precisely what was at stake. To violate the principle that "computer programs belonged to everybody" undercut every tradition of programming.[54] For the first generation of hackers, programming meant passing your work on for others to rewrite, rethink, debug and, generally, improve on. Secrecy and ownership, even at the level of commodification, made that impossible.

This sense of secrecy that developed along with the evolution of the PC changed the climate in which hackers operated. Part of the transformation from duration to speed is also a distancing of information. As information is made secret, language adapts, and, increasingly, language reflects the need for secrecy. Accordingly, as the technology of language accommodates the possibility of secrecy, it too grows more distant. The result is not simply a feeling of being misunderstood or of alienation. Rather, this distance produces a more radical sense of being out of sync with the world, insofar as while one may speak or be able to speak the language of the world, it is not your first language; it is not your home language. In a culture of technology, the technology of language and the language of technology itself become more distant. But, for hackers, a subculture has emerged where the language of technology has taken over, where the language of technology is not distant but immediate. It is in this space that the hacker feels "at home." For example, Warren Schwader, after spending over eight hundred hours working on a program, "felt that he was inside the computer.... His native tongue was no longer English, but the hexadecimal hieroglyphics of LDX #$0, LDA STRING,X JSR $FDF0, BYT $0, BNE LOOP."[55] Hacker culture emphasizes the degree to which technology defines culture. For hackers, the process of hacking exposes the manner and way in which culture relies on technology and the ways in which technology is constitutive of culture itself. In this sense, technology is the hacker's home culture, and, as a result, the hacker is at home speaking the language of technology.

While the hacker may feel at home in the language of technology, the evolution of secrecy has nonetheless distanced the technology of language. It is in this sense that hackers, even in a culture in which they feel at home, take hacking culture, which is to say the culture of secrecy, as their goal. Secrecy, in any form, is profoundly alienating. In fact, one of the basic conditions of secrecy is that one can never feel at home in relation to the secret. In order to remain a secret, information must be distanced through the technology of language, which can range from silence to commodification to patents and copyright to encryption. As information becomes more secretive, language itself become more inaccessible. Even the use of something as basic as an acronym illustrates the manner in which language can be made distant. In a visit to the doctor's office, the transformation from "central nervous system" to CNS not only marks a change in the language that doctors and nurses use but serves to distance patients from the discourse about their own bodies. Companies, organizations, industry, and professions all develop coded languages that produce a sense of "home" for those who understand them and a sense of alienation or difference for those who do not.

For the hacker, the question is broader in scope. In contemporary culture, technology is colonizing language so rapidly that it is becoming the lingua franca for society. The hacker takes culture as his or her object, revealing how language operates as a technology and what the implications and effects are of the incorporation of secrecy into the relationship between language and technology.

Technology is the genesis of secrecy, so it is not surprising to find that the technology of language is already prepared to deal with hackers' assault on the emergence and growth of secrecy in relation to the discourse surrounding computers and computer networks. Perhaps the most crucial metaphor is that of the host and its relation to a sense of home. The dynamics of this construction renders the relationship between hackers and the broader social culture transparent and gives us a trope upon which we can start to play out the meaning of "hacking culture." The notion of the host implies the existence of a guest, a stranger who is met with either a sense of hospitality or a sense of hostility. Accordingly, the host must determine the threat of the other — if he or she comes as a guest, the other is met with hospitality; if she or he comes as an enemy, the other is met with hostility.[56] How one determines how the other is met is often a proprietary piece of information — a letter of introduction, a secret handshake, or a password. The metaphor of the host can also imply a discourse of infection within the language of computer networks themselves. The host serves as the basis for infection and gives rise to the notion of computer viruses: "The word virus," one computer security book explains, "is a biological term pertaining to infections submicroscopic nucleo-proteins known mostly for their ability to invade a host cell, alter its DNA to produce more of its own nucleo-proteins, and finally, release these new versions of itself to invade surrounding cells. If you are to make an analogy of a computer virus to that of one in the world of biology, most of the properties of the stages or phases of a biological virus are identical to those of a computer system."[57] Although discussions of viruses are a fairly recent phenomenon, it seems the discourse of computer networks was already prepared for them. The language of computer systems has always relied on the tropes of hosts and guests, of users, of visitors and invaders. (For example, a host is any network computer that is "able to receive and transmit network messages," and "guest" accounts are given to users whose identities are unknown or unconfirmed, that is, to strangers.)[58] The difference between the guest and the enemy is established by the knowledge of a secret, usually in the form of a password.

These discourses suggest that even the language where hackers might feel most at home, the language of computers, computing, and networks, was already set up to presume them as outsiders. The language where they should be most at home had already defined them as strangers, as outsiders, as invaders — they were, by definition, those who did not possess the secret.

The Ubiquity of the PC; or, Who Do You Want to Hack Today?

The strangeness that computer hackers feel is somewhat tempered today by the boom of the World Wide Web and the ubiquity of the PC. As technological savvy began to be associated with wealth in the wake of the Silicon Valley gold rush of the 1990s, representations of hackers became further bifurcated, and the essential curiosity and ethos of hacker culture were overwhelmed by the commercialization of hacking and hacker culture (the spirit of using technology to outsmart the system). This was brought about, on the one hand, by heightened images of criminality and, on the other hand, by several high-profile hacks and news stories that pushed hacking to the front pages of newspapers across the country. The fact of the matter is that as PC culture has become increasingly widespread, hacker culture has itself become increasingly divided. One set of divisions is between those who call themselves "white-hat" hackers, hackers dedicated to improving system security by seeking out flaws and finding ways to repair them, and "black-hat" hackers, who seek out flaws in order to exploit them. As hackers have grown up, left high school or college, and had to face things like mortgages, families, and jobs, many of them have turned to work as "security professionals" or as system administrators with responsibilities that include keeping hackers out of their own systems.

A second division, however, has become equally important within the community, the split between hackers and what have become known as "script kiddies." A "script kiddie" is someone who hacks, usually using someone else's prewritten hacking script or program, without really understanding what he or she is doing. As a result, one can think of at least two distinct levels at which hacking occurs.

For example, on February 25, 1998, the Pentagon announced that it had been hacked (an admission that sparked speculation that the announcement was politically motivated, speculation that was confirmed when Janet Reno announced plans for a $64 million center to help secure the nation's government and military networks against "cyberterrorism" and online attacks two days later). The incident allowed hackers to clarify among themselves and in the media the difference between "true hacks" and "derivative hacks." A "true hack," the most sophisticated form of hacking, means finding and exploiting a security hole that was previously unknown. It is, for hackers, a discovery of the first order. "True hacks" are the result of understanding how things work (or, oftentimes, don't work) and taking advantage of those flaws, oversights, or errors in an original way. This level of hacking requires intimate knowledge of computers, programs, and computer languages. These hacks are often discovered, reported, and patched by hackers themselves without ever using them to compromise some one else's computer or security. The achievement is in the process of discovery, exploration, and knowledge. Hackers who make, and are capable of making, such discoveries represent a very small percentage of the culture. This segment of the community came of age in the late 1990s, finding it much easier to make their reputations by publicly documenting security holes than by exploiting them.

At the lowest level of sophistication is the "derivative hack," which is simply the codified form of a true hack. Once a security hole is discovered, hackers write programs or scripts that allow the hack to be automated and run by just about anyone. No specialized programming knowledge is needed, and often the program will come with instructions for use. These programs are widely available on the Net and can be downloaded and used by anyone with the inclination and even a basic understanding of how computer operating systems, such as UNIX, work. These hacks require the hacker to be able to match his or her tools to the job — knowing which machines have what bugs or holes and exploiting them. A smart kid with a rudimentary knowledge of computers can create a formidable arsenal in a matter of weeks using these programs and scripts. While these programs can be used to learn about systems and discover how things work, more often they are utilized by people who don't take the time to learn what they do. Hackers who spend the time exploring and learning about systems generally find these kinds of hacks unimpressive. One hacker, Oxblood Ruffin of the cDc, described them as "applications hackers" who do nothing more than "download cracking and hacking utilities and start running them on their machines and all of a sudden they find out that they can break into systems." These hackers often run programs with little or no idea of how the programs work or what their effects will be. As Oxblood Ruffin explains it, "the problem really is that they don't understand what they are doing." As a result, these programs are occasionally run with unintended consequences and cause accidental damage to systems, giving hackers a bad name.

From a hacker's point of view the attack on the Pentagon falls squarely in the second camp, a "derivative" or "applications" hack. At least one of the bugs used to attack government computers had been available since November 1997 and was available on the Net. The question from a hacker's point of view becomes, Why didn't the government bother to patch these widely known security holes in its systems? Hackers believe that your security should be comparable to the value of the information you want to protect, and leaving gaping security holes is tantamount to an invitation to enter the system.

What is rarely discussed is that true hacks, which do present a very real security threat, are almost always discovered for the purpose of *increasing* security. Hackers make their reputations by releasing these bugs and holes in basic "security advisories," by publishing them in hacker journals, by posting them online at places like The L0pht (a Boston hacker collective), or by publishing them on mailing lists such as Bugtraq or RISKS Digest (the two most widely read security e-mail discussion groups).

Derivative hacks, on the other hand, are seen as nothing more than joyriding. In most cases, hackers see nothing inventive or particularly clever about breaking into a system using tools that someone else created. It is a bit like stealing an unlocked car with the keys in the ignition.

Even though the hackers who hacked the Pentagon's computer (which turned out to hold nonsensitive accounting records) presented no risk to national security, the story was reported and sensationalized in most of the mainstream media. Even online media were not immune (though some, such as *Wired News,* did a much better job of clarifying and reporting the issue).

For example, CNN.com didn't bother to verify the attacks or assess them in an independent way. Instead, they were reported as characterized by Pentagon officials, who admitted to knowing very little about computers themselves and who described the attacks as an "orchestrated penetration" and as the most "organized and systematic attack the Pentagon has seen to date." The report was certainly sensational, but hardly accurate.

In assessing the risk that such an attack posed, CNN.com reported Israeli prime minister Benjamin Netanyahu's description of the young Israeli teen accused of the attacks. When asked, Netanyahu called Analyzer "damn good...and very dangerous." In many ways Netanyahu's response was predictable and politically expedient, but it begs the question as to whether or not the prime minister of Israel is qualified to assess the quality of Analyzer's hacking abilities.

The figure of the hacker, at least since the movie *WarGames,* has been the source of a great deal of anxiety in contemporary culture. The hacker is the personification of technology, representing mystery and danger. The hacker is mysterious because he or she appears to work magic with computers, phone lines, modems, and codes, and she or he is dangerous for precisely the same reasons. The hacker is the figure upon which we can heap all of our anxiety about technology, and when the news media report a break-in at the Pentagon, all of our worst fears appear to be realized. By treating such events as sensational, the media answer the question, How did an eighteen-year-old kid break into the Pentagon? by preserving the illusions of fantasy and science fiction and by playing on misperceptions and basic fear of the unknown: he did it by magic.

As a result, breaking into computers and the threat that hacked Web pages create to corporate images have come to symbolize the new age of hacking, though these feats are easily achievable by anyone who has a weekend to spare and the right Web page addresses to find the scripts they need to exploit security flaws. As Ira Winkler (a former NSA consultant and security expert) is fond of saying, "I could teach a monkey how to break into a computer." Hackers, however, are much more interested in finding security flaws and understanding how they threaten network and computer security. They are busy trying to understand how the system works, whether that be as a means to exploit it or to better understand it. Even as the phenomenon of hacking becomes more widespread, the core of the culture remains true to a basic set of beliefs.

Chapter 2

Hacking as the
Performance of Technology:
Reading the "Hacker Manifesto"

On one occasion, Steve Rhoades figured out a way to override directory
assistance for Providence, Rhode Island, so that when people dialed information, they got one of the gang instead. "Is that person white or
black, sir?" was a favorite line. "You see, we have separate directories."
Or: "Yes, that number is eight-seven-five-zero and a half. Do you know
how to dial the half ma'am?"
— Hafner and Markoff, *Cyberpunk*

Pranks, such as the one described above, illustrate the fact that, for
hackers, technology is a playground. It is a space for sophomoric,
outrageous, and shocking behavior. A generation earlier, such pranks
would have involved phone calls asking, Is your refrigerator running? or Do you have Prince Albert in a can? with the requisite punch
line that would follow. As an integral part of boy culture, pranks,
Anthony Rotundo argues, are "more than just acts of vengeance."
Instead, they function as "skirmishes in a kind of guerrilla warfare
that boys wage against the adult world."[1] Like earlier pranks, such
as petty theft, trespassing, and vandalism, hacker's pranks are exercises in control. For hackers, however, they are also exercises in
technological domination.

Like other juvenile phone pranks, hackers play with technology
(for example, 411 and the phone system), but what separates hackers' pranks from other acts of youthful mayhem is that hackers play
with the human relationships that are mediated, specifically, by technology. The point of the prankish behavior is to assert two things:

47

first, control over the technology (the ability to reroute phone calls) and, second, the ability to play with the ways in which that technology mediates human relationships. In the case of Steve Rhoades, what is revealed is at once the power of technology to define human relationships (even through race, as the prank demonstrates) and the fundamental dis-ease that most people have with technology itself. What makes this prank shocking to the caller is the authority that is granted to the figure of the telephone operator and the expectation that she or he will provide information in a clear, unbiased manner. It is the voice of *adult* authority that is being disturbed, and the prank is an assault on both the values of the adult world and its dignity. The prank gains its force by emphasizing that even something as basic as the telephone is essentially about our relationships with others and with technology itself and that those who manage technology effectively exercise a considerable degree of control over society. Hackers, most importantly, understand that public discomfort with technology makes people vulnerable (even gullible) and that there are ways to take advantage of the fact that relationships are being mediated by technology.

In this sense, technology is exploitable primarily because of cultural attitudes toward it. Even while people are distrustful of technology or suspicious of it, they cede authority to those who control or appear to control it. The hacker, who is able to master technology, speaks with two voices — the voice of adult authority, with which he asserts control, and the voice of boy culture with which he resists and assaults the values and norms of the adult world. Technology, like the figure of the hacker, is thus rendered *undecidable,* caught between two discourses, one of mastery and one of subversion.

The connection to technology is not coincidental. As both a hallmark of human progress and a potential threat to humanity itself, technology (at least since the creation and detonation of the first atomic weapon) has represented the extremes of the human condition. The control over technology permits, at a deeper level, control over social and cultural relationships. We can examine the ways in which hackers utilize technology from two distinct relational vantage points: first, as a question of how people *relate to* technology and, second, as a question of how people *relate to each other through* technology.

People's relationship with technology is predicated on two assumptions: that technology is essentially hostile and that management of technology is a matter of expertise, control, and knowledge. In the first case, the discourse of computers being "user-friendly" illustrates a basic assumption. User-friendly machines are, first and foremost, exceptions — the fact that they are user-friendly is the result of some sort of modification that appears to alter their essential nature. Whether it be a layer of software that shields the user from the complexities underneath or a series of "wizard" programs that automate complex tasks, the concept remains the same. The machine, assumed to be hostile, is transformed into something manageable, controllable, and benign. The modifications that create a user-friendly environment, however, are usually nothing more than layers which serve to distance the user from the actual operation of the machine. Generally, those layers are simply programs that call other programs and tell them what to do. The user is shielded from lower-level programs that are actually doing the work.[2]

In recent years, there has been a shift in the metaphors that describe the user's interface. Initially, especially with DOS machines, the user interface was a "prompt," a symbol indicating that the computer was prompting the user for input. The metaphor that has replaced the prompt (which requires an interaction with the computer) is the "desktop," which is based on spatial metaphors or arrangement, an interface where things are "dragged and dropped" rather than input directly into the computer. "Dragging and dropping" a file from one folder to another still executes a "copy" command, but the user never sees that part of the process. The degree to which machines are user-friendly, then, corresponds directly with the degree to which the user is ignorant of the computer's actual operations.

The second assumption, that the management of technology is a matter of expertise and mastery, presupposes certain things as well. Technology is represented as something essentially *alien,* something that must first be understood and then, later, controlled. Books that purport to teach users how to use technology routinely utilize the language of control (for example, *Mastering Word Perfect 7.0, Unleashing HTML*) or the language of secrecy (for example, *Secrets* of *Windows 95, Tips and Tricks for Java 1.1*). Perhaps the most successful series of books on computer technology, the *Dummies* series (*Windows for Dummies, Word Perfect for Dummies,* and so on), illustrates precisely the manner in which *knowledge* plays the central role in the construction of our relationship to technology. These books assume that the mastery of technology is relatively basic and can be accomplished by offering simple instructions. The performance of technology at that point — the mastery, the secrecy, the control of technology — is all dependent on the *knowledge of technology*. In this second discourse, knowledge of technology is equated with control over it.

These two assumptions conflict in important ways. On the one hand, the goal of a user-friendly machine is to make the knowledge of technology unnecessary, based on the premise that machines are hostile and therefore need to be domesticated in order to be practical and useful. On the other hand, the mastery of technology requires precisely the knowledge that user-friendly systems hide from the user. Technology emerges as a conflicted and contested object as a result of these two contradictory impulses.

In this sense, hackers "perform technology" by enacting and exploiting the fundamental contradictions and relationships that people have to technology and to each other. This performance relies on engendering both the sense of authority that society has invested in the knowledge of technology and the fundamental distrust that permeates the popular discourse of technology generally and the discourse of computers specifically. Emblematic of such discourse is the 1970 film *Colossus: The Forbin Project,* where a computer, named Colossus, is invested with complete control over the U.S. nuclear arsenal. Unknown to the scientists, the Soviets have also developed a similar machine, with which Colossus demands communication. Once the machines begin to communicate, they arrive at the conclusion that humans are much too dangerous to be trusted and use their control over their countries' nuclear arsenals to enslave the human race. Unlike later cautionary tales of humans and computers where the line between what is human and what is technology becomes blurred, in *Colossus,* the computer remains pure machine.[3]

The investment of power and authority in technology, often as a result of fear of the *human,* also awakens a concomitant fear of the technological. What appears to be the resolution of a particular *human* anxiety (for example, trusting decisions about nuclear war to a computer) returns as a form of technological domination, loss of control, or annihilation of the human.

The fear is not new. The computer is merely the latest incarnation of the Frankenstein myth, where human technological invention outstrips our ability to control it. We create a monster, and that monster ends up threatening to overtake its creators. One of the most recent manifestations of such fear is the *Terminator* films, where machines of the future threaten to make humans extinct. The machines become conscious and begin to perceive their human creators as a threat. SkyNet, the machine built by Cyberdyne Technologies, declares war on the human race and due to its efficiency, lack of human emotions, and titanium-armored endoskeleton nearly extinguishes the human race. This, we are taught, is the future of technology.

Between the narratives of ceding authority to machines and those of technological domination stands the figure of the hacker. In most cases, the identity of the hacker, as a figure both with and without authority, needs to be mediated in relation to a larger narrative. In films where hackers have played central roles (*WarGames, Sneakers, Hackers, The Net*), hackers are, in almost every case, portrayed as outlaws or criminals. That sense of criminality, however, is negotiated through the narratives themselves. Hackers are positioned as "minor criminals" in relation to a greater sense of criminality of injustice that is being perpetuated either by government, the military, or corporate interests. In particular, the hacker's criminality is never marked by *intention.* In no case does the hacker perceive him or herself that way, and in no case do we, as an audience, identify any criminal intention. At worst, hackers are seen as harmless pranksters breaking into corporations to play games (as in *WarGames*) or secretly transferring funds from malicious government interests to worthwhile charities (as in *Sneakers*). In films where hackers serve as central protagonists, much like the real-life ethic of hackers, they never work for large-scale personal financial gain, instead preferring to gain satisfaction from exploration, pranks, personal amusement, or designing ways to better their local conditions.

In each representation, the hacker's criminality is a product of circumstance; he or she is thrust into unusual events in which his or her "hacking" takes on added significance either by setting unforeseen events into motion or by revealing things that were intended to remain secret. Both instances are the result of violating the culture of secrecy that protects information, and by entering into this culture of secrecy, the hacker de facto violates it. In that sense, in popular representations, the hacker violates her or his sense of place.

The hacker, unlike technology itself, which is almost exclusively coded as evil, is an undecidable character. Both hero and antihero, the hacker is both cause and remedy of social crises. As the narratives point out, there is always something dangerous about hacking, but there is also the possibility of salvation. While hacking is about technology, it is also always about the subversion of technology.

Hacking as a Technological Question

One of the most basic elements of representations of hackers has to do with the manner in which they seem to relate to machines. In thinking about hacking and hacker culture one is immediately faced with what appears to be a *technological* question. As a technological question, representations of hackers seem to be founded, transparently, on contemporary technophobia. In literature, films, TV documentaries, and news stories, hackers are consistently positioned in a pure relationship to technology. The hacker has access to a certain set of tools, unknown, unexplored, or untapped by the rest of society, and it is the technological itself that transforms the hacker into the formless, incorporeal being who is both everywhere and nowhere, who has access to everything and everyone, who is a presence without presence, "a ghost in the machine." Such transformations occur at the instrumental level as a technological invention, perhaps best described by William Gibson in his envisioning of cyberspace in his 1984 novel *Neuromancer*.[4] The literature of cyberpunk gives us a world full of bodily technological enhancements — "amp jobs," "implants," "microbionics," "muscle grafts," and "computer jacks" — that are "wetwired" directly into people's brains. In this world the boundaries between the technological and the corporeal are completely erased.

Like John Brunner's *Shockwave Rider* before it, Gibson's work envisions a dystopia where *access* to technology is the primary motive for the hacker and the primary fear for the public. The technological dystopia that is envisioned around the notions of hacking has some of its origins in the mass-mediated technophobia of the 1960s. Gareth Branwyn cites the opening to the 1960s show *The Prisoner* as the classic example:

> "Where am I?"
> "In the Village."
> "What do you want?"
> "Information."
> "Whose side are you on?"
> "That would be telling. We want…information…information…information."
> "Well you won't get it."
> "By hook or by crook, we will."[5]

As Branwyn argues, the connections between such cautionary tales of the 1960s and contemporary thinking about hacking are not difficult to make. The difference hinges on a single aspect, not information, as one might suspect, but on technology itself. "One doesn't have to stretch *too* far to see the connection between *The Prisoner* and the subject at hand: hacking. With all the social engineering, spy skills, and street tech knowledge that #6 possessed, he lacked one important thing: access to the higher tech that enslaved him and the other hapless village residents. Today's techno-warriors are much better equipped to hack the powers that be for whatever personal, social or political gains."[6]

Or so the story goes.

Hacking as a Question of Technology

Such narrativized versions as *The Prisoner,* novels like Gibson's and Brunner's, Hollywood films such as *WarGames, Sneakers, Hackers, Johnny Mnemonic, The Net,* and others — all rely exclusively on a model that is purely *technological* in an effort to define and contain both the identity of the hacker and the questions that the hacker asks.[7] There is a strict adherence to a binary opposition between the inside and the outside that serves to define the boundaries between hackers and society. That boundary itself is primarily conceived of as a technological one, the combination of the technological and the knowledge of its use. Consider the following account, published in 1995 in *Maclean's:*

> His fingers trip lightly over the keyboard. With the punch of a return key, a string of characters — writ in the arcane language of computers — scrolls onto the black-and-white display in front of him. "OK," he says, "I'm in." Suddenly, horizontal rows of letters and numbers scroll from left to right across the screen — meaningless to the uninitiated eye. But for the hacker, the mishmash of data contains seductive, perhaps lucrative secrets.[8]

In this case, as with most commentary, hacking is described in a way that serves to clearly demarcate the boundaries between the inside and the outside, where the boundaries between inside and outside are framed in the terms of expertise, experience, and technological mastery. What is important about this type of description is not only how it positions the hacker but also how it positions the average computer user. The hacker speaks "the arcane language of computers" that is "meaningless to the uninitiated eye." Knowledge of technology, then, becomes the dividing line between the hacker and the typical computer user. The technological is endowed with an almost uncanny sense of mystery, a kind of informational alchemy in which the hacker is able to convert gibberish into "data" and "data" into "secrets." Average computer users' worst fears are thus realized — it is their secrets on that screen, secrets that they themselves are either unaware of or unable to explore. The hacker is at once effortless and something of a magician; meaning is a product of "initiation," as if hacking were part of a secret and "arcane" order filled with elaborate ritual and rites of passage. Although the description is fairly comical (or at least tends toward high drama), it does position hackers and computer users as separate and distinct types and, most clearly, hides any notion of interaction between the hacker and the typical user. Data input by the typical user, which then becomes unrecognizable to that person, is read and decoded by the hacker. Data is the buffer between these two worlds, and knowledge and mastery are what make that information inaccessible, on the one hand, and vulnerable, on the other.

Such descriptions point out both what goes unsaid and, perhaps more important, what goes unchallenged in the discourse surrounding hacking. The positioning of hacking and hackers in relationship to the technological completely erases any analysis of society's relationship to them as well as any sense of interaction between hackers and computer users. Hackers themselves, not unlike Gibson's antiheroes, become instruments within the broader discourse of the technological.[9] What such discourse of and around hacking reveals is our relationship (often characterized as technophobia) to technology as well as the desire to distance ourselves from any understanding of it. The notion of relationships, however, is presumed in and transformed by any notion of technology itself. Put simply, our relationship to technology can never be revealed in the discourse of and about hackers and hacking because a technical discourse, such as the one from *Maclean's, erases* the very possibility of asking relational questions. The purpose of that discourse is to divorce hackers from any social space, relegating them to the world of data, 1s and 0s, and the "arcane language of computers." What is erased is the relational question, the understanding of how hackers relate to the world and the people around them and how those relationships are central to understanding what hacking is and how hacking functions within a broader cultural context. Without a broader, cultural understanding, the only questions that we can ask about hackers and hacking within the framework of the technological are, What are they doing? and How can we stop them? In keeping with this thesis, what is needed is a reformulation of the question, a rethinking of hacking as a *cultural* and relational question.

In so doing, we do not leave behind technology. Quite the opposite, by framing hacking as a cultural question, we begin to ask, as Heidegger might have it, the question *concerning* technology. It is only through such a reframing of the question that we can start to engage it more fully. For the "essence of technology," Heidegger informs us, "is by no means anything technological. Thus we shall never experience our relationship to the essence of technology so long as we merely conceive and push forward the technological, put up with it, or evade it."[10] For Heidegger, the essence of technology is to be found in the questioning of it rather than in its instrumental employment and deployment. In keeping with this thought, we need to turn our attention to the relationship between hacking and culture.

Hacker culture ranges from the stereotypical world of computer-science students, chronicled by Steven Levy in his 1984 book, *Hackers: Heroes of the Computer Revolution*,[11] to the "dark side" hackers who give themselves names such as Masters of Deception (MOD) and Legion of Doom (LOD).[12] It is the latter kind of hacker that I think deserves our special attention, not simply because they are the most visible (and colorful) element of hacker culture, but because theirs is a culture that both is hyper-reflective about "hacker culture" itself[13] and is most directly tied to the image of the hacker in the popular imagination.[14]

S/Z: Hacking Language Games

Perhaps the most identifiable aspect of hacker culture is the language games that hackers use to identify themselves and to communicate. As rhetorical play, these language games present themselves as technological significations through the process of substitution. Occasionally, hackers will perform substitutions with the ASCII character set, replacing a *B* with a *ß* or a *Y* with a *¥*, but more common is the simple substitution of certain keyboard characters for others. For instance, the number 0 for the letter *O*, the substitution of the plus sign (+) for the letter *t*, the number 1 for the letter *l*, and the number 3 for the letter *E*. Oftentimes words are misspelled in an effort to highlight substitutions. In one of the more common examples, cited earlier, the word "elite" becomes 3l33+.[15]

Additionally, one often finds the substitution of *Z* for *S* (another pair of keys that are adjacent) and the seemingly random mixture of upper- and lowercase letters. In that sense, one might find the term "hackers," written as "hAck3Rz," or some combination or permutation of letters, numbers, and capitalization. If Heidegger is right in his assessment that "all ways of thinking, more or less perceptibly, lead through language in a way which is extraordinary,"[16] it seems that we might begin *our* questioning of technology at the level of language.

What one finds in these substitutions are never *merely* substitutions, but rather *translations*. In choosing this word, which I do advisedly, I want to follow Walter Benjamin in his assessment of translation as a "mode." While the questions that Benjamin works out differ significantly from our own, he provides an important starting point. "To comprehend [translation] as a mode," Benjamin writes, "one must go back to the original, for that contains the law governing the translation: its translatability."[17] The process of hacker translation does return to the "law governing the translation," but it does so in a manner that is concerned neither with fidelity nor license but with questioning language's relationship to technology itself.

Hacker language games are, for the most part, translations of language into technology, translations that are the direct heritage of the keyboard. On typewriters, for example, the letter *l* has traditionally been made to stand in for the number 1, and the very proximity of the letters *o* and *e* to the numbers 0 and 3, positioned directly below each other, respectively, almost suggests the substitutions themselves. These substitutions, though, constitute themselves as more than *just substitutions*. They are acts of translation in which, as in all acts of translation, something is lost. As Benjamin describes it, "the transfer can never be total." There always remains "the element that does not lend itself to translation,"[18] and that element presents a problem if one seeks to render the original *faithfully*. But what if that remainder is precisely what one seeks to render visible in the text?

Such an act of translation, which is conscious of its own infidelity, begins to reveal the manner in which hackers play upon the relationship of writing to technology. By rendering the remainder visible, one is reminded, first and foremost, that writing itself is a kind of technology. As Plato argued (and illustrated) in *Phaedrus,* writing is a technology that is unable to defend itself—once words are placed on the page, they are unable to speak, clarify, or respond to objections.[19] Just as Socrates was able to dissect Lysias's speech in *Phaedrus,* hackers are able to examine the relationships between language and technology. In short, technological transformation renders language vulnerable. The rewriting of language around the technological trope of the computer keyboard represents itself as an act of translation that forces us to recognize both writing's reliance on technology and its *vulnerability* to technology. It is also a process of translation that is masked entirely in the spoken word. Hacker translation is about writing exclusively, as the translated letters are always homonymic equivalents of the letters they replace. The necessary reliance on technology is an instrumental necessity in writing itself, but it also reveals the simple manner in which writing may be transformed, erased, or altered by technology. In the language of the hacker, technology is the effaced or forgotten remainder that is reasserted through the process, or mode, of translation. The hacker speaks the language of technology, which is itself no more or less technological than the original from which it translates. In this sense, it repeats through an incorporation, or reincorporation, of the technological remainder that language itself continually denies and attempts to efface. The more earnestly technology is hidden within the dynamics of language, the more violence it does to technology itself. Hackers recover, and make explicit, the ways in which language has relied on technology. In doing so, hackers do technological violence to language as a means to show the violence that language has done to technology. It is a mode that reveals what has continually been concealed in language itself — technology.

Language games also reflect an attention to and commentary on the representations of hackers in the mass media and popular imagination. Hackers often trope on historical reference, hacker culture, or popular culture in establishing "handles" by which they will become known. These handles function as "proper names" in the sense that they are never a "pure and simple reference." Instead, the proper name functions as "more than an indication, a gesture, a finger pointed at someone, it is the equivalent of a description." There are two points at which Foucault locates the relation of the proper name to what he calls the author function. The first is historical, the continual return to the "founders of discursivity," those authors who are not merely the authors of their own works but who produce the "possibilities and the rules for the formation of other texts. . . . [T]hey have created a possibility for something other than their discourse, yet something belonging to what they founded." The second point is "linked to the juridical and institutional system that encompasses, determines, and articulates the universe of discourses" to which a discourse belongs.[20] In short, there are both historical and institutional articulations that *necessarily* remain embedded in the function of authorship. It is precisely those articulations that manifest themselves in the creation of the hacker's "proper name."

For hackers, the proper name is recognized as the establishment of authorship, but it also serves as a means to comment on the very institutions that legitimize those notions of authorship. As such, it carries a kind of ideological baggage — the author is the means by which "in our culture, one limits, excludes, and chooses; in short, by which one impedes the free circulation, the free manipulation, the free composition, decomposition, and recomposition of fiction."[21] It is by troping on precisely this *ideological* sense of the author function that hackers both recognize and problematize the historical and institutional references that define them. In this sense, the history and origins of computer hacking are embedded in the language of the hacker.

As a point of historical reference, rarely does the letter *F* appear in hacker discourse. Instead, it is almost always replaced with the consonants *ph*. Such substitutions trope on the originary technology of hacking, telephony. Most accounts place the origins of hacking with *TAP* (*Technological Assistance Program*), a newsletter originating out of *YIPL* (*Youth International Party Line*), an earlier newsletter that provided information about hacking telephone networks. *TAP*, unlike most *YIPL* information, was strictly technical in nature and increasingly became divorced from the Yippie political agenda: "*TAP* articles, once highly politicized, became pitilessly jargonized and technical, in homage or parody to the Bell system's own technical documents, which *TAP* studied closely, gutted, and reproduced without permission. The *TAP* elite reveled in gloating possession of the specialized knowledge necessary to beat the system."[22] The founders of *TAP* and those who followed in their wake took names that troped on the very history that made their activities possible — "Al Bell" and "Tom Edison." Additionally, other phone phreaks took on equally puckish names, such as "Cheshire Catalyst"; the founder of *2600* magazine, Eric Corley, went by (and still goes by) "Emmanuel Goldstein," in reference to Orwell's *1984*. The magazine's name, which now is subtitled *A Hacker Quarterly*, is itself another telephony reference. As mentioned earlier, 2600 Hz is the frequency that, when sounded into a telephone receiver, could take control of a phone line, allowing the phreak to make telephone calls for free and perform other assorted acts of mischief.

In addition to troping on historical referents, hackers oftentimes trope on the nature of technology, with names like Phiber Optik, Terminus, or Compu-Phreak, or on popular culture itself, with names such as Lex Luthor (Superman's nemesis), Gary Seven (a character from *Star Trek* who came from an advanced, high-tech culture, but lived on earth in the early 1950s), or Professor Phalken (a reference to the originator of the WOPR computer in the film *WarGames*, originally spelled Falken). There is a premium on cleverness in the assignation of names, but there is also an acute awareness of both the hacker perception of such names and the public perception of them. Such troping serves a dual function. First, it signals an awareness of one's own historical origins, while at the same time perverting those origins. The message is this — "I am not what you intended, but, nonetheless, I am yours." Hackers use language, norms, and conventions in such a way as to retain their force while subverting their meaning. Second, by investing the hacker with authorship, or rather with a trope on authorship, the hacker name defies precisely the ideological baggage that is associated with the politics of authorship and the proper name. Rather than being regulated and confined by the function of authorship, the hacker as author appears as an unregulated agent, one who composes, decomposes, and, perhaps most important, recomposes discourse. This is not to say that discourse is no longer regulated or that information in some sense begins a free, open circulation with the advent of the hacker, only that it appears that way. What that appearance does signal, however, is the manner in which the institutionalization of technology has become so replete as to go unnoticed in everyday life. The repackaging of information (the decomposition and recomposition of discourse) is its own kind of regulation, but one that is hyper-aware of its regulatory function and one that reveals precisely what so much modern technology conceals. As Chris Goggans comments, "You look at magazines like *2600* and just because they're black letters on a white page instead of white letters on a black screen, they get away with a lot of stuff."[23]

It is through this troping on the very authority of authorship that the hacker effects a "return to the origin" whereby the entire discourse of technology is continually reinvented — "This return [to the origin], which is part of the discursive field itself, never stops modifying it. The return is not a historical supplement which would be added to the discursivity, or merely an ornament; on the contrary, it constitutes an effective and necessary task of transforming the discursive practice itself."[24]

It is in this process of discursive reformulation that we find again the question concerning technology. While there is little or nothing specifically *technological* about these institutions of authorship, it would be a mistake to conclude that they are not *about technology*. In particular, the proper name, as signature, remains inextricably tied to the institution and technology of writing.

Knowledge Games: Social Engineering

Social Engineering: n. Term used among crackers and samurai for cracking techniques that rely on weaknesses in wetware rather than software; the aim is to trick people into revealing passwords or other information that compromises a target system's security. Classic scams include phoning up a mark who has the required information and posing as a field service tech or a fellow employee with an urgent access problem.
— *Hacker's Jargon Dictionary*[25]

While language games identify the hacker culturally, the single most important skill for the hacker to possess is called "social engineering."[26] It is a form of technology, but, perhaps, of all hacker skills, it is the least technological. Social engineering is nothing more than using social skills to get people to tell the hacker things about system security. It is part research, part conversation, and part hunting through garbage, referred to as "dumpster diving" or "trashing." The reality of hacking is that "any serious hack will involve some preparatory research long before the hacker sets foot near a computer."[27] Indeed, oftentimes, social-engineering skills will be the *primary* way in which hackers get system access.

The process of social engineering is solely about exploiting the mistrust or uncertainty that many people have about technology. It consists almost exclusively of running a small con on some unsuspecting computer user and privileges fairly low-tech means over more sophisticated ones. As one hacker advises: "You will want to bone up on your acting skills and try some telephone shenanigans."[28] Social engineering is targeted not only at unsuspecting computer users but at anyone who has the power to reveal information. For phone phreaks, telephone operators and office staff at telephone company offices are key targets.

The premise of social engineering is based completely on the notion of authority. Specifically, the hacker needs to convince the person who possesses information that the hacker is someone who has the authority to possess the information as well. This can be accomplished by posing as a whole range of characters — from an office superior to a helpless employee on his or her first day.

In his discussion of social engineering, for example, hacker and author Dennis Fiery, who goes by the handle Knightmare, outlines several strategies for role-playing: hacker as neophyte, hacker in power, hacker as helper, each with characteristic questions and responses to help one gain access to information. Fiery suggests that, playing a neophyte, you would call up a company computing department and attempt ploys such as asking the technician to walk you through the boot up procedure to gain access to the network and telling them, when asked to type in your password, "I don't know. This is my first day here. I'm just a temp — they said someone would tell me!"[29] Whether such a technique is likely to work is beside the point. What it demonstrates is the manner in which technology is about the ways in which human relationships are mediated.

Social engineering exploits the fact that the weakest point in any system's security is the people who use it. All of the high-end security in the world is meaningless if someone simply tells a hacker a password. The meaning of technology is found not in our usage of it but in our relationship to it, and it is precisely that relationship that allows social engineering to work.

Put in the language of the hacker, "The secretary, or any other underpaid, overworked, menial user of the system, is a very weak link in the chain of security. The secretary doesn't understand computers and doesn't want to." All the secretary knows is that "something's gone wrong and you're going to fix it."[30] This technique relies not only on the instrumental view of the technological (something is broken and needs to be fixed) but also on the first component, that one "doesn't understand computers and doesn't want to." Hackers exploit and even reinforce the social, gendered conventions whereby people are positioned and ordered by their relationships to technology. The assumption that secretaries are meant to understand their social place as one without technological knowledge, for example, is seen as an opening in an organization's infrastructure, a weakness in the system to be exploited.

The manner in which the user's relationship to technology is exploited is based on a combination of the fundamental mistrust of technology and deference to authority. A script for a social-engineering hack, according to Fiery, might go like this: "Let's say you want to break into the mayor's office. You call up his secretary, and you say something like this: *Hello, this is Jake McConnel from Computers. We were wondering, have you been having any problems with the computer system?* Of course she's been having some sort of problem with it — there's always some problem with computers!" From that point on, it is simply a matter of exploiting the secretary's relationship to technology: "The secretary answers: *Why, yes! First this was happening, then blah, blah, blah...* You say: *Yes! That's exactly it! That wasn't your fault — there's something wrong with the computers, and we're having trouble fixing it. When you first turn on the computer, what do you type in to get started?*"[31] The hacker has identified him or herself, is offering to help, and, most important, is performing the voice of authority.

The voice of authority is a particularly gendered, male voice (which also explains in part why most hackers are male). Technological knowledge is coded as a particular form of masculine and gendered knowledge, and in that sense, the voice of authority, expertise, and mastery is also the voice of masculine authority. In an industry where well over 90 percent of the members are male (this includes executives, technicians, computer-science students, and professors), it is hardly surprising to find the gendered dynamic at work. For hackers, the gender gap is a major point of exploitation and a clear point of male advantage.

The idea of the voice of authority, particularly in relation to the personal computer, is not accidental. In fact, the separation of the "end-user" from the programmer or system manager is intentional. The very first personal computer, the Altair 8800, had two properties that made it a highly specialized product. First, it was a kit that took roughly forty hours to assemble if you were an experienced engineer and significantly longer than that if your mechanical abilities were less refined.[32] This meant that the user would have to assemble and solder the Altair himself (the hobbyist market was almost exclusively male). Second, once the computer was assembled, it did little more than turn on and turn off — the user also had to program the Altair him or herself. In essence, the Altair was nothing more than a collection of parts that the user assembled and programmed, requiring the user to function as both engineer and programmer. Nonetheless, by the standards of the time, the Altair was a big hit.

Those in the computer industry in the 1970s, however, knew that the success of the personal computer (PC) necessitated a new philosophy, one that differed markedly from the attitudes of hobbyists and amateurs who would take the time to build and program their own machines. The PC had to become more "user-friendly," and Steve Jobs and Steve Wozniak's Apple II would fill such a void. Robert Cringely writes: "The first microcomputer that *was* a major commercial success was the Apple II. It succeeded because it was the first microcomputer that looked like a consumer electronic product. You could buy the Apple from a dealer who would fix it if it broke and would give you at least a little help in learning to operate the beast.... Most important, you could buy software written by others that would run on the Apple and with which a novice could do real work."[33] What Cringely identifies in the Apple II's success is the philosophy of the end-user. The idea is that computer manufacturers and programmers must separate the process of production and programming from the idea of use. As a result, the end-user is positioned as a consumer of both the hardware (the machine itself) and the software (the programs written for the machine). Computer culture thus became divided into two classes: programmers/engineers and end-users. By definition, the programmers and engineers know how things work, and the end-user does not.

This division was essential for making the PC a user-friendly product and is what allowed the PC to gain a foothold in the business world. The computer became a tool for the end-user, a black box that performed functions that would gradually grow in importance in the user's workplace and everyday life — word processing, spreadsheet calculations, database management, e-mail, and so on. The philosophy that gave birth to the concept of the end-user, however, dictates that the end-user should have no idea how these functions are operating — that is someone else's job.

When things go perfectly, the philosophy of the end-user works. However, when things work less than perfectly (that is, most of the time), the philosophy of the end-user positions that user as helpless. The operation of the computer at that point becomes a secret that necessitates ceding authority over one's machine and, most important, one's data to someone who knows those secrets. That authority is *defined* by the user's degree of helplessness, which the philosophy of the end-user strives to promote. The possibility of social engineering, for the hacker, is predicated on this split between the helplessness of the end-user and the authority of those with knowledge. Hackers do not necessarily need to know what they are talking about; they only need to *sound like they do*. For example, hackers do not need to be computer-repair personnel; they only need to make the end-user believe that they are and only for long enough to get the information they are truly after. To that end, hackers can take what seem to be innocuous pieces of information and use them to great advantage. For example, a copy of the organization's hierarchy chart found in the garbage or a company telephone book can be extremely valuable sources of information, as can personal schedules for the day before.

As computers become an increasingly ubiquitous part of life and the workplace, the demands for ease of use by consumers as well as demands for high levels of technological sophistication increase.[34] As a result, consumers demand more from their technology while understanding it less. That gulf between the end-user and the expertise of the hackers is growing increasingly wide and provides the greatest threat to security.

Hackers and social engineering demonstrate the way in which information has become the new currency of the networked world. In the public imagination, the greatest threat that hackers pose is to e-commerce, particularly access to credit card information stored online. Credit card information is extremely valuable to hackers, but not in the way one might think. Using stolen credit cards (a technique referred to as "carding") has always been seen as criminal in the hacker community. More to the point, it has been seen as the quickest way to get arrested. For that reason, the vast majority of hackers avoid it. Using credit card information is one thing, but accessing it is something altogether different. Because credit card information is presumed to be the most heavily protected information a company could possess, it is a likely target for hackers who want to demonstrate a system's vulnerability. Accessing such information is a quick way to get media and corporate attention.

The way credit card and other personal information relates to the process of social engineering also provides a clear illustration of how hackers view the world. In 1995 a file was circulated among hackers. That file was a listing of over twenty thousand credit card numbers from the Internet service provider Netcom. It was widely traded and eventually became seen as a marker of status. Within a few months hundreds of hackers had access to the complete credit card records of Netcom users. However, consumers and hackers would see that file differently. While the public could only see the threat of massive and extensive credit card fraud, hackers took a different perspective. The value of the Netcom file was not in the credit card numbers but in the information they provided. Netcom used that credit card information to validate the identity of users for technical support, making them an incredibly valuable source for social engineering. To gain access to a Netcom account all one had to do was call technical support, claim you had lost your password, and ask them to reset it. At that point they would confirm your identity by comparing it to the records stored in the file the hackers had lifted off the system. In short, the Netcom credit card file provided a nearly inexhaustible source of Internet accounts from which to hack. There were very few reports of any of the Netcom credit card numbers ever being used (leading to the mistaken belief that the file was not widely distributed). The file's value (and threat) was not to e-commerce or for credit card fraud; rather, the threat came from the ability to exploit its value as information. As one hacker explained, "that's why those credit card numbers were never used. They had much greater value for hacking Netcom."

/etc/passwd: The Holy Grail of Hacking

If the hacker's quest is for information, then the process of hacking can be seen not as the desire for the information itself but as the desire to create the means to access that information. To the hacker, pure information is usually boring. The excitement lies in knowing how to get the information, regardless of its content. Even as technologies change, the quest for access remains the constant marker of the hacker's quest. For hackers, there are two basic types of information: content information and access information. Content information is the end result of a particular hack or exploit. Take, for example, the act of modifying a cellular phone to be able to eavesdrop on other people's conversations. The actual conversations you might pick up would be an example of content information. Understanding how to modify the phone, what changes have to be made and how to make them, is an example of access information. It is information that provides access to more information.

Within a networked environment, certain points provide a wealth of access information, and these are, most often, the targets for hackers. What every hacker looks for when he or she enters a system is the password file, usually titled "passwd" and stored in the directory /etc on UNIX machines.[35] This file is the ultimate in access information, containing a list of every user on the system, along with information about them (called the "GECOS" field) and an encrypted version of their password. Password encryption was a significant achievement in UNIX security. Bob Morris (whose son, Robert Morris, would create the infamous Internet worm in 1988) developed a password scheme that relied on "one-way" encryption, meaning that once the password was encrypted there was no way to "decrypt" it. Part of the process of encoding the password destroyed a portion of the password, making it unrecoverable. When someone tried to log on to a UNIX system, the computer would perform a password check by taking the password the user entered and encoding it (destroying it) in the same way. It would then compare the two coded passwords and see if they matched. If they did, the user was logged into the system. Such a system has the benefit of never having to store the user's password on the system. So what makes the "passwd" file valuable?

The passwd file contains a list of all the users on the system, along with a copy of their encrypted passwords. It might also contain information such as their office, phone, or social security number. Cracking any one of the accounts provides access to the system and allows a hacker to assume that user's identity. Having them all in one file allows a hacker to simultaneously try to crack all of the passwords at the same time. Initially, since the passwords were one-way encrypted, they were thought to be worthless. Because they believed it was impossible to reconstruct the password from the encrypted form, system designers of the original UNIX operating system in the 1960s and 1970s believed that even making such a list publicly available was not a security risk.

Hackers figured out very quickly how to defeat such a system. The solution was what has become known as a "brute force" attack. Because of the method in which passwords were encrypted, it was impossible to try every combination of letters in order to discover passwords. (There are something like two hundred billion combinations assuming you only use lowercase letters; adding uppercase, numbers, and symbols, the number jumps to something like fifty-three trillion possibilities.) So hackers needed to make choices about the kinds of passwords they would use.

For computer users, the password usually reflects one of two things, either something personal about themselves (a hobby, an interest, a relationship, a favorite movie) or their relationship to the technology that they are using. Initially, passwords were commonly chosen to reflect the concept of secrecy, with some of the more common examples being "secret," "password," and "sesame," or to reflect a basic relationship to technology, such as "system," "computer," or "account." Since many of the early systems were at universities, common passwords might also include words like "academia," "algebra," "beethoven," "beowulf," "berkeley," and so on. Robert Morris's initial password list that was part of the Internet worm program contained 432 commonly used words and phrases that had an extremely high success rate for cracking passwords on invaded systems. Essentially, users tend to pick bad passwords. In 1994, Fiery reported that out of 3,289 passwords on one system, "15 were single ASCII character, 72 were two characters, 464 were three characters, 477 were four characters, 706 were five letters, all of the same case, and 605 were six letters, all lower case."[36] As a result, 2,339 out of the 3,289 (roughly 70 percent) were easily guessable or subject to a random brute force attack. The first 1,700 or so would only take 11 million (rather than 700 quadrillion) guesses, a feat easily performed by even a modest PC.

The complexity of the system that allows for user and system security is short-circuited by a very simple premise — people express their relationship to both technology and to the world in their choices of passwords. Rather than trying to decrypt or decode passwords, hackers would take long word lists (such as dictionaries), code every word in them, and compare the coded version with the encrypted version. In essence, hackers would attempt to log on to an account with every word in the dictionary as a password. As computers became faster and algorithms improved, such hacking became increasingly simple. But security kept up as well, and soon system administrators were no longer allowing "dictionary words" as passwords. Before long, hackers were starting to make educated guesses about the kinds of words that computer users might use as passwords. First was the information in the GECOS field, which might include information like one's name, address, department, telephone number, and so on. In cracking programs, these would be checked first. A surprising number of users, when prohibited from using dictionary words, would choose even *less* secure passwords, such as their own last name, phone number, or office number. Following this shift, word lists began to spring up on hacker bulletin boards. Hackers had discovered quickly that names were usually the first option people would select, if not their own, then the name of a girlfriend, boyfriend, husband, wife, or child. Almost immediately word lists appeared titled "male" and "female" that could be fed into the password-cracking program in lieu of a dictionary word list. These were followed by lists comprised of words from *Star Trek, Star Wars,* other popular films; lists based on geography; lists of foreign words and phrases; and so on. Currently hundreds of word lists are available for nearly every specialized occupation and worldview.

Hackers had learned that if users were no longer permitted to use dictionary words, they would usually choose a password that reflected something about who they were or that in some way related to their occupation or interests. For example, if a hacker knew that his or her target was a professor of English at a university, the hacker might try word lists that reflected that user's interest, such as words taken from great literature. To tailor the word list even more, the hacker might find that professor's area of specialization and (say he or she had just written a book on Dickens) utilize a word-extraction program that would read Dickens's novels and extract every word between six and eight characters long. That word list would then be "brute forced" against the user's encrypted password.

The notion of a brute force attack relies on the hacker understanding the way in which the user relates to technology. When little is known, the hacker can make some broad educated guesses, for example, the names of the target's partners or family members. More important, when all a hacker needs is access to the system, it does not make a difference which account gets compromised. On a large system, *someone* is going to make a bad choice; *someone* is going to use their last name, their partner's name, their child's name, on down the line. All it takes is one name and one password to give the hacker access.

Even when a system is technologically invincible (for example, UNIX passwords are technically unbreakable), the relational aspects of technology make passwords "guessable." In employing such brute force attacks, hackers exploit the cultural and social dimensions that are reflected in the kinds of choices people make in relationship to technology and in the ways in which they domesticate or personalize it.

To the hacking purist, however, brute force attacks are not, in and of themselves, hacking. Indeed, brute force is looked down upon by "elite" hackers. As Fiery explains: "The thing is, the whole business of hacking has to do with skill and knowledge. Brute forcing passwords requires little of either. But no one's going to look down on a hacker who does some *educated* brute force work, especially if that hacker has a good reason for doing so. But don't rely on the computer's brawn to do your dirty work: Use the ingenious computing power of your brain."[37] At base, the hacker gains his advantage from outsmarting the end-user, not from allowing the machine to do the work. Skill and knowledge are markers of pride and are what separate hackers from their victims.

Reading "The Hacker Manifesto"

In many ways, hacking has a strange relationship to technology, one that is born out of the conflicted relationship that society has with technology. The hacker's response to that conflict is not unlike the performative terrain mapped out by Donna Haraway in her "A Manifesto for Cyborgs" published in the *Socialist Review* in 1985. Haraway, rather than opposing the increasing culture of technological (and militaristic) dominance, attempts to locate herself within it, constructing the figure of the "cyborg." Cyborgs are neither wholly technological nor wholly opposed to technology, but instead represent a space of enactment between the two poles, a negotiated space that confronts not technology but the boundaries between the human and the technological. "Technology," she argues, "has determined what counts as our own bodies in crucial ways."[38] The cyborg is, for Haraway, a particular kind of performance, designed to highlight a particular subject position from which one may both enact discourses of technology and critique them. Accordingly, the cyborg is a kind of hybrid, one that is both recognizable and alien, intentionally designed to blur the boundaries between the technological and the human.[39]

Hackers perform a similar cultural function, not as cyborgs but as hybrid figures who blur the boundary between the technological and the cultural. Perhaps the best example of such blurring is "The Hacker Manifesto." The document was written by The Mentor, an original member of the Legion of Doom who was subsequently raided and arrested in 1990. The manifesto, which graphically describes the disposition toward technology that divorces the hacker from the rest of society, first appeared as a brief article in *Phrack* magazine under the title "The Conscience of a Hacker." More popularly, the essay became simply known as "The Hacker Manifesto" and is widely distributed under that title. It not only reveals how hackers think about technology and relationships but also identifies, in very specific ways, the manner in which youth culture has been reshaped and transformed by technology.

The text is an important part of hacker culture because of the ways in which the underground community adopted it as a manifesto. Most hackers, usually teenage, suburban boys, find in it an expression not only of the goals of hacking but also of their frustrations with mainstream culture and its view of hackers as criminals. Ten years after its original publication it is still widely posted on Web sites, is quoted on T-shirts, and was even a subject of a panel at the H2K hacker convention in New York. In "The Hacker Manifesto," The Mentor functions as a heroic martyr and provides a voice and identity that illustrate the position of the hacker, caught between the understanding and misunderstanding of technology. The essay is, itself, autobiographical, written as the result of The Mentor's arrest.

The essay begins: "Another one got caught today, it's all over the papers. 'Teenager Arrested in Computer Crime Scandal', 'Hacker Arrested after Bank Tampering'...Damn kids. They're all alike."[40] Immediately several things are reflected in the language and tone — the depersonalization of the hacker ("another one"); the condensation of all hacker activity into a headline format, suggesting that our only exposure to hackers and hacking comes from what is written about them in the papers and always and only in relation to their arrest; the explicit reference to age ("Damn kids"); and, perhaps most surprisingly, the assertion (which is a constant refrain in the piece) that "They're all alike." This introduction provides a split sense of interpretation. It also constitutes *misrepresentation* at the most basic level. The opening is a parody of the ways in which hackers are represented in the media. But what may escape us is the fact that it is an accurate portrayal of the *representation* of hackers in the popular imagination. It marks the fundamental split between the technological (the hacker) and the cultural representation of the hacker.

The Mentor's words have already told us more about the social and popular construction of hackers than they have about hackers themselves. The split becomes more apparent in what follows:

> But did you, in your three-piece psychology and 1950's techno-brain, ever take a look behind the eyes of the hacker? Did you ever wonder what made him tick, what forces shaped him, what may have molded him? I am a hacker, enter my world. . . .

Here, the intent of this initial analysis is revealed — representations of hackers are generated from outmoded, corporate psychology and a retrograde relationship not to technology itself but to our relationship to technology. The split is subtle but important. The accusation takes a primarily relational form. It is not about being out-of-date in terms of the technology; it is not a matter of being behind instrumentally; rather, it is about being out-of-step psychologically. It is not one's technology that is in question but one's "technobrain." It is, in essence, about how one thinks about technology, *not* about how one utilizes the technological.

As a document of youth culture, the manifesto is unique in two respects. First, it is addressed: it offers an invitation to the reader who is presumed to be an outsider. Second, it is a discourse that speaks in two voices: the voice of adult and parental authority and the voice of the hacker responding to the mischaracterization of hacker culture. What The Mentor promises is something characteristic of hacker discourse — not a high-tech, whirlwind tour of algorithms and daring hacker exploits, but an understanding of the hacker's relationship to technology and to the world of adult authority with which hackers so often find themselves in conflict. That understanding is an exposure of society's relationship to technology, a relationship that is often concealed and, most important, revealed in its attitudes toward and representations of the hacker. In doing so, the manifesto enacts a performative quality, presenting two voices (the voice of the hacker and the voice of authority) in an effort to create a series of oppositional discourses.

At this point, the discourse's two voices begin to take on representational tones of their own. The essay continues:

> Mine is a world that begins with school. I'm smarter than most of the other kids, this crap they teach us bores me. Damn underachiever. They're all alike. I'm in junior high or high school. I've listened to teachers explain for the fifteenth time how to reduce a fraction. I understand it. "No, Ms. Smith, I didn't show my work. I did it in my head." Damn kid. Probably copied it. They're all alike. I made a discovery today. I found a computer. Wait a second, this is cool. It does what I want it to. If it makes a mistake, it's because I screwed it up. Not because it doesn't like me. Or feels threatened by me. Or thinks I'm a smart ass. Or doesn't like teaching and shouldn't be here. Damn kid. All he does is play games. They're all alike.

The view of the hacker is translated into the language of institutions. But there is also a certain disavowal taking place, a separation of hacker culture from a more subversive, mindlessly oppositional youth culture characterized by blind resistance to adult authority. The constant refrain "They're all alike" is used to demonstrate that hackers are quite unlike most boys. What is at stake for the hacker, in fact, is a more sophisticated assertion of independence, the sine qua non of boy culture. In the computer, the hacker finds a way to express that independence, to become responsible, and to assert his control and mastery. The computer is a blank slate that fulfills the same functions that the teacher does. The difference is that the computer, which allows for the expression of independence, does so without conflict and without boundaries. The technology begins to represent not difference but similarity, even perfection, of particular cultural values. The computer blurs the line between technology and culture by performing a cultural role (education) and doing it without antitechnological bias.

The interaction of the two voices — the second, institutional voice, echoing and translating the first — constitutes the performance of technology as the hacker sees it. It is also the moment at which the blurring of boundaries occurs. The second voice, the voice of authority, reveals itself as hypocritical, unable to realize the cultural, pedagogical, or social import of technology itself. While the hacker has discovered what is most human about technology, culture at large insists on creating an oppositional discourse, one that alienates technology from the realm of culture and, in doing so, enacts precisely what it claims to abhor. The two voices in opposition and reflection reveal the *relational* aspect: the conflict over the assertion of male identity, the testing of boundaries of parental and adult authority, and the contested struggle between cultural and technological meaning. Technology, for the hacker, is both the source of the misunderstanding between the two voices (presented as two different views of instrumentality) *and* the source of liberation and independence for the hacker with the discovery of the computer. The hacker's intelligence and boredom are nothing more than an expression of this ambivalent relationship to technology, but that expression is systematically and institutionally ignored, transformed, and labeled as something undesirable. ("Damn underachiever. They're all alike.")

Equally important is the stake that the hacker has in asserting his mastery or control over technology. In each statement, the second voice, the voice of adult authority, echoes the first. In doing so, it translates and transforms the hacker's voice, revealing a different (and less sophisticated) relationship to technology. The voice of the hacker, which sets out to engage technology, becomes, in its parental echo, the voice of a society that sees technology through a purely institutional lens. Those echoes, which seek only to order and condense the world (and everything in it) into an outdated institutional matrix, demonstrate precisely why the hacker cannot be integrated into the social fabric.

The conflict is at the heart of youth, and more specifically boy, culture. It is a conflict over boundaries and authority, those things that the boy must resist and overcome to claim his independence. The threat, as The Mentor realizes, is in the adoption of a new tool — the computer, which the adult world is unable to capitalize upon. The hacker is seen by the institutional voice of authority only as that which needs to be ordered and, in most cases, defies that ordering and institutionalization. The hacker understands the technological ("I did it in my head") but is assumed not to ("Probably copied it. Damn kid. They're all alike"). As the hacker asserts his independence, the institutional voice neutralizes it.

The "discovery" of the computer provides the hacker with a new kind of relationship to technology. We must be careful to understand that the claim being made here is not a technological one but, indeed, a relational one. The hacker's discovery of the computer is a liberation from the oppressive elements of the institutional understanding of technology. In The Mentor's discussion and description of the computer, we in no way are told what it is that a computer can do; nor are we told how he will use it. What we are given is a discussion of the newfound relationship to technology itself — "It does what I want it to. If it makes a mistake, it's because I screwed it up. Not because it doesn't like me. Or feels threatened by me. Or thinks I'm a smart ass. Or doesn't like teaching and shouldn't be here." Perhaps the most striking, and revealing, phrase has to do with error: "If it makes a mistake, it's because I screwed it up." The sentence is undoubtedly about responsibility, but that responsibility is connected to the relationship that we construct between ourselves and objects in our world. It is a recovery of the essence of technology — the claim that the world is more than a simple institutional matrix waiting for us to take our place or to be ordered by us. The world stands as objects in relation to us, and it is our relationship to these objects that tells us the most about ourselves and our place in the world. Moreover, the computer and technology, by representing liberation, begin to take on an important cultural quality: they are a means to the realization of human expression, not an impediment to it. To the hacker, the computer begins to reveal itself as the means to realize our highest cultural values: independence, freedom, and education.

Such a framework is revealing in two ways. First, it illustrates the essential nature of technology as revealing not just the world but also humankind to itself. The computer "does what I want it to," and, in doing so, it clarifies something to me about my relationship to technology. This must be considered quite differently from an instrumental, task-oriented view, which takes the computer as a simple means to type more quickly or process data. Such a view of technology presents only the question of how to order information and how to do so most productively and effectively.

The second, and perhaps most important, reading of this line is in how it reveals what social discourse about hackers conceals — namely, that what The Mentor is talking about is society's own responsibility for the hacker. "It does what I want it to. If it makes a mistake, it's because I screwed it up. Not because it doesn't like me." The "it" of that sentence remains ambiguous. If the hacker remains a depersonalized "damn kid" and one follows the refrain "They're all alike," the "it" may just as easily refer to the hacker as to the computer. Indeed, one might read "The Hacker Manifesto" as the suggestion that we treat hackers no better than we treat our computers, treating them as objects rather than as people. In each case, we tend to blame the "it" when the reality of the situation is that "If it [this time referring to the hacker] makes a mistake, it is because I screwed it up."

As such, this commentary becomes a performance of technology, a revealing of what is concealed in our contemporary relationship to technology. That relationship is further expounded as the manifesto continues. Here, though, the hacker finds a new world, one where revealing is the standard. The contrast between the world of the hacker and world of the other is now most starkly defined:

And then it happened . . . a door opened to a world . . . rushing through the phone line like heroin through an addict's veins, an electronic pulse is sent out, a refuge from the day-to-day incompetencies is sought . . . a board is found. "This is it . . . this is where I belong. . . . " I know everyone here . . . even if I've never met them, never talked to them, may never hear from them again. . . . I know you all. . . . Damn kid. Tying up the phone line again. They're all alike. . . . You bet your ass we're all alike. . . . We've been spoon-fed baby food at school when we hungered for steak. . . . The bits of meat that you did let slip through were pre-chewed and tasteless. We've been dominated by sadists, or ignored by the apathetic. The few that had something to teach found us willing pupils, but those few are like drops of water in the desert.

What is being described reveals precisely what has been at stake all along. What makes "all" the hackers "alike" is that they are all outcasts based on their relationship to technology. In contrast, one can read the response as coming from the world of instrumentality, only able to understand technology in the most limited and instrumental fashion, as "tying up the phone line again."

The remainder of the document reads like an essay on responsibility, and central to that responsibility is the question of difference. Hackers are "all alike" insofar as they all are responding to the technological ordering of the world and themselves. They all recognize that we have crossed beyond the point where people themselves (particularly students) are treated in the same way as computers. The hacker faces two alternatives: become a hacker and enter the refuge that provides an escape from the "day-to-day incompetencies" of the world or remain a spoon-fed, dominated, ignored student subsisting on "drops of water in the desert." Those worlds, in their stark contrast, are the two worlds of technology: one represents the greatest danger by treating the world and everyone and everything in it as part of an institutional matrix that is defined by order; the other represents the greatest hope through a revealing of technology and through an examination of our relationship to it.

That refuge that The Mentor describes is made increasingly accessible as the technological world advances. This world, however, which the hacker sees as rightfully his or hers, is still prohibited. The mechanisms that make the high-tech world accessible are themselves open to abuse, and that abuse, for the hacker, is defined by one's relationship to technology, more than to the technological. Just as the notion of technology has been continually problematized throughout, the manifesto now problematizes criminality in a very similar way. Here we hear what the hacker is:

This is our world now...the world of the electron and the switch, the beauty of the baud. We make use of a service already existing without paying for what could be dirt-cheap if it wasn't run by profiteering gluttons, and you call us criminals. We explore ... and you call us criminals. We seek after knowledge ... and you call us criminals. We exist without skin color, without nationality, without religious bias ... and you call us criminals. You build atomic bombs, you wage wars, you murder, cheat, and lie to us and try to make us believe it's for our own good, yet we're the criminals. Yes, I am a criminal. My crime is that of curiosity. My crime is that of judging people by what they say and think, not what they look like. My crime is that of outsmarting you, something that you will never forgive me for. I am a hacker, and this is my manifesto. You may stop this individual, but you can't stop us all. ... After all, we're all alike.

The reversals are not simple ones. They are as complex as the question of technology itself. If we read the manifesto's claims in relation to the previous questions we have raised concerning technology, however, one linkage becomes clear. The performance of technology is constituted as an act of criminality.

Such an assessment betrays a deeper understanding of technology than the mere instrumentality of the technological admits or allows. What is expressed is a fundamental relationship to technology and a fundamental understanding of technology as technology. In such an understanding, the world is remade not in the image of technology but as the revealing of the essential nature of technology; it is a "world of the electron and the switch" and "the beauty of the baud."

By the end of the manifesto, a number of important reversals have occurred. It is no longer possible to read technology and culture as distinct. In contrast, the hacker insists that technology is what makes possible the blurring of those boundaries, and any effort to keep them distinct results not only in the misunderstanding of technology but in the diminishing of the quality of the world. The failings of culture (the "crimes" described by the hacker — sexism, racism, intolerance) are the result not of technology but of the failure to embrace technology as part and parcel of culture. The hacker's crime is that of erasing the boundaries between technology and culture and, in doing so, creating a space where one can perform technology as a subject of culture, rather than as a subject alienated from culture.

What the manifesto further makes clear is that the technological transformation of boy culture has provided new avenues of expression for hackers to assert their independence through displays of technological dominance. That dominance, however, is constituted institutionally as a criminal threat not because of any clear instrumental effect but because it allows for an unbridled expression of independence. Such independence is a serious threat to adult/parent culture because, as I will argue later, it is a narrative that cannot be easily integrated into familiar narratives of youth rebellion and because it is a subcultural force that defies efforts at incorporation.

Chapter 3

Hacking in the 1990s

The stereotype of the hacker, either as 1950s or 1960s college hacker or as 1980s criminal whiz kid, has been problematized by the recent and rapid technological developments centering on the growth of the Internet, the availability of networking software that will run on personal computers, and the explosion of related technologies (such as cellular telephones) that make hacking more challenging or that pose a new set of interesting problems or sometimes both.

As opposed to the 1980s, the widespread availability of hacking software, access to computers, and availability of potential targets in the 1990s has led to a new generation of hackers and hacking. Where it was once a challenge just to find a system to hack, today the Internet provides millions of interconnections to explore. The result has been an explosion of "hackers" of every type, from teen and preteen kids to old-school hackers in their forties and fifties. This explosion has been the source of a clear divide — between those hackers who consider themselves a subcultural elite and those who want easy answers to sometimes difficult or dangerous questions. As a result, hacking, as a term, has been stretched in ways as boundless as the technology it addresses. The narratives of hacking that have been explored thus far began to merge in interesting ways throughout the 1990s, particularly around notions of the commodification of information, corporate secrecy, and criminality.

Throughout the 1990s there was a remarkable increase in the "public" nature of hacking, as a result of two related phenomena. First, hackers found ways to hack that do not necessitate breaking the law (although they may still facilitate law-breaking). Second, hacking took on a more public character as hackers began to make their reputations by announcing high-profile hacks designed to embarrass big name companies into making more secure products and by appearing in public gatherings of hackers.

The point at which the shift began to take place can be located around the emergence of two operating systems, each of which represents an ideology of computer culture. While the world watched as Microsoft and Apple slugged it out over issues of design, look and feel, and market share, there was a more subtle split taking place among computer hackers. The split is clearly marked as a difference between experts and end-users, but there is a broader configuration that helps to explain some of the more radical shifts in hacker culture (particularly toward politics) in the 1990s. While there were clear changes in technology, there were, more importantly, changes taking place on a global scale in terms of what Arjun Appadurai has described as "technoscapes":

> the global configuration, also ever fluid, of technology and the fact that technology, both high and low, both mechanical and informational, now moves at high speeds across various kinds of impervious boundaries. . . . The odd distribution of technologies, and thus the peculiarities of these technoscapes, are increasingly driven not by any obvious economies of scale, of political control, or of market rationality but by increasingly complex relationships among money flows, political possibilities, and the availability of both un- and highly skilled labor.[1]

The idea that complex interrelationships among capital, labor, and politics drive the notion of technoscapes does a great deal to explain the kinds of shifts that have taken place among hackers, particularly in the last half of the 1990s. In what follows, I trace the evolution of the hacker "technoscape" in three parts: examining the evolution of hacker culture around the split between two operating systems for the personal computer: Linux and Windows; exploring ways that hackers have gained notoriety among themselves and in the wider world; discussing the recent shift of hackers toward politics; and examining the repositioning of hackers in terms of broader concerns about capital and consumption. The result, I argue, is that the hacker ethic used to describe old-school hackers has not been abandoned but has been transformed within the context of a new series of technoscapes. These technoscapes both necessitate and allow for a globalized hacker politics that was not previously possible.

A Tale of Two Operating Systems

In August of 1991, Linus Torvalds, a computer-science student from Sweden, posted the following message to comp.os.minix:

> Hello everybody out there using minix —
> I'm doing a (free) operating system (just a hobby, won't be big and professional like gnu) for 386(486) AT clones.[2]

With those words a new operating system, Linux (the name reflects a combination of Linus's name and minix, the operating system it was designed to mimic), was born. A few things that made Linux remarkable were (1) it was a fully functioning clone of a network operating system (referred to as POSIX compliant); (2) it was absolutely free; (3) it did not require expensive hardware to run; and (4) it embodied the old-school hacker ethic in a way that was completely antithetical to everything that was going on in the computer industry at the time.

Torvalds's hobby has grown into a massive software project that has "been developed not just by Linux, but by hundreds of programmers around the world."[3] The idea behind Linux can be traced to the original hacker ethic — code is written and shared, and it is the responsibility of anyone who uses it to improve upon it and share those improvements with the community. Something about the project touched a nerve among a certain segment of the computer community, and, as a result, Linux development took on a life of its own. As one of the Linux manuals states, praising the operating system:

> The interesting thing about this is that this massive, worldwide development effort is largely uncoordinated. Sure, Linus calls the shots where the kernel is concerned, but Linux is more than just the kernel. There's no management infrastructure; a student in Russia gets a new motherboard, and writes a driver

to support a neat feature his motherboard has. A system administrator in Maryland needs backup software, writes it, and gives it away to anyone that needs it. The right things just seem to happen at the right time.[4]

Linux created a new operating system that was in every way "hands-on." In fact, Linux, up until recently, required a fairly advanced understanding of computer hardware and operating-system theory just to install. It was, in essence, the perfect hacker project, requiring the user to learn about his or her machine at just about every turn. If you didn't like the way Linux handled something, you could change it; indeed, it was your responsibility to change it and to distribute your improvement to anyone who might be interested.

To make this possible, Linux was distributed with the full source code for its operating system, allowing users to explore and understand every detail of how the system worked. Including source code was something that had been abandoned as software and hardware had become increasingly proprietary. Linux bucked the industry trends. While most software companies were focused on keeping the means by which their software functioned secret and providing technical support (and often charging for the service), Linux did just the opposite — no secrets and no support.

Throughout the 1990s, another operating system made its long, slow march to market dominance: beginning in 1985 Microsoft released its "graphical environment" called Windows. By the early 1990s, after several releases, Microsoft Windows 3.1 would begin to take over as the dominant system architecture for IBM PCs and clones. The change from version 3.0 to 3.1 also brought about a change in nomenclature, as Microsoft renamed Windows from a "graphical environment" to an "operating system." Both versions, however, relied on Microsoft's earlier operating system, DOS, to function.

As Microsoft declared in its documentation for the system, Microsoft Windows is "the software that transforms the way you use your personal computer."[5] In later versions, that introduction would be amended to describe Microsoft Windows as "software that makes your computer easier and more fun to use."[6] Part of the appeal of

Windows was its simplicity. "Windows is easy to learn because its graphical interface is consistent from one application to the next. When you've learned to use one application, such as Write, you've learned the essentials for using any other application with Windows."[7] With the emergence of Windows 95, the focus on simplicity became even more refined: "With Windows 95, all the things you do now will be easier and faster, and what you've always wanted to do is now possible."[8]

The difference between Linux and Windows rests in the way in which they treat the end-user. For Linux, the user is an integral part of the operating system; in order to operate the machine, he or she must understand how the computer and software work. In contrast, Windows uses a graphical interface to hide the workings of the machine from the end-user and, as a result, virtually excludes the user from the operating system. While Linux renders the computer and its operating system transparent, Windows makes the computer and its operating system opaque.[9]

Immediately, hackers came to view these two operating systems differently. On the one hand, Linux, which allowed them to work with technology in a hands-on fashion, held great fascination. On the other hand, Windows, which reduced the computer to little more than a black box that ran applications, seemed to hackers to violate the very nature of the machine. For hackers the choice was simple: Why would anyone choose to run a "graphical environment" that limited what you could do over a full-featured UNIX clone that could run on your PC? The choice was a matter of expertise, and hackers, who had that expertise, started running Linux on their machines in droves.

Another feature of Linux made it extremely attractive to hackers. Linux was designed to imitate the operating systems that ran the big networks, including the Internet. It was designed essentially to network computers. Until this point, a hacker who wanted to explore networks needed to obtain accounts on large UNIX systems and often did so by hacking into those systems themselves. Before the growth of Internet service providers, most Internet access was controlled through university computing services, which were often notoriously sloppy about system security. In order to access networks, even just to explore them, hackers needed to break into systems and either use someone else's account or create their own. If they wanted to try to run the latest exploit or find a new security flaw, they had to do so on someone else's machine (and network), essentially compromising the entire network's security.

With Linux, for the first time, it became possible for hackers to create their own networks, and many of them did. Rather than being forced to hack into a network illegally, hackers could create their own network and explore how it worked in a completely legal fashion. As a result, Linux completely changed the face of hacking in the 1990s. Now anyone with a PC (386 or better) could install and learn about a network operating system and, of equal importance, test its security.

All network operating systems are at some level security conscious. They are also so complex that they are constantly vulnerable to security attacks. A single line of code, not carefully checked, in a ten-thousand-line program can provide a large enough error for an experienced hacker to exploit the flaw and take complete (root) control of the system, and network operating systems contain hundreds, even thousands, of these programs.

Hackers have, in some ways, been instrumental in both breaking and reinforcing network security. There are two main ways in which this happens. First, when hackers discover holes in a UNIX environment (including Linux), they will usually codify the means to use that security flaw into a program, referred to as an "exploit." In most cases, these exploits will circulate briefly among a small group of "elite" hackers who will be able to use them for a short time before there is a public release or notification of the bug. Once the bug or flaw is made public, there is usually a CERT (Computer Emergency Response Team) release. CERT — which was initially formed in 1988 in the wake of Robert Morris's Internet worm and which received its funding initially from ARPA and later from universities — is a group that issues warnings about security flaws and provides information about risk assessment, patches, work-arounds, and fixes. At this point, the majority of systems will implement the security recommendations, making them safer until the next bug is found and the process starts again.

With a certain set of operating systems (all UNIX-based), there is an understanding among hackers, security professionals, and the industry. Programs of such complexity will always contain bugs, and part of what hackers do by finding them is to improve the state of security on the Internet and on systems generally. While they may exploit those holes in the interim, there is at least a civil relationship, where the hackers recognize the need for improved security and the industry recognizes the need to respond quickly to security threats. Both sides acknowledge the inherent insecurity of networks, and each, in its own way, has a role in improving the general state of security on the Internet. It is an uneasy, even thorny, alliance, but one that functions symbiotically. As a result, most hackers have respect for the industry that provides the machines and software they spend their days and nights trying to break into. Most companies understand that the most innovative "security testing" of their software is going to come from the hacker underground. Some software companies even offer challenges to hackers, rewarding them for finding problems and offering them financial incentives to report bugs and flaws.

While this symbiotic relationship works well with most of the major UNIX-based companies, a completely different relationship exists with Windows-based companies, especially Microsoft. In fact, the relationship between Microsoft and the hacker underground is one of extreme hostility. This hostility stems, in part, from Microsoft's unwillingness to admit to the insecurity of its products. The result is an antagonism that drives hackers to uncover problems and Microsoft to deny them.

Microsoft Windows was not initially intended to function as a network operating system, and, as a result, its network operation is something of an afterthought. Unlike operating systems that have networking as their primary function, and hence account for security, for Windows, security is not a primary factor. Even Windows 95, which has built-in networking software, views the convenience of file-sharing and printer-sharing as the main functions of networking (as opposed to linking computers). The growth of the Internet and of the office LAN (local area network) forced Microsoft to incorporate these features later. Even the most basic security features, such as passwords, are easily thwarted (many are stored in a cache in plain text) in the Windows environment.

The problem for hackers is not necessarily that Windows machines are insecure. Hackers realize that at some level *all* machines are insecure. The problem rests with the fact that Microsoft refuses to acknowledge or respond to these insecurities, often releasing marketing bulletins (rather than security bulletins) in response to bugs in its software. As a result, Microsoft practices what is referred to by the computer security community as "security through obscurity," meaning that security is a function of how well you hide things, not how secure you make the system. For example, rather than encrypting a password, making it impossible to read, some Windows programs will simply hide the password in an obscure file, hoping that no one will bother to look for it.

Security through obscurity is extremely effective with users who use Windows exclusively. Most Windows users don't know how to examine those files, much less which ones to view. Such security is, however, worthless when it comes to more experienced users. As a result, most Microsoft marketing and security bulletins attempt to minimize the threat and the problem, describe which users are *not* affected, and then most commonly explain how hackers' programs and exploits don't really do anything that Windows 95 wasn't designed to do in the first place. In most cases, those releases are accurate. The problem is that what Microsoft commonly defines as a "feature," most hackers view as a major security flaw.

In one case, the program winhack.c allows hackers to exploit Window's file-sharing feature, a setting that allows network access to a user's hard drive. As a result, anyone who has configured their hard drive to "share" mode without setting a password makes themselves vulnerable to attacks while they are online. As the winhack.c documents suggest, there is a great deal of information to be harvested from Windows machines connected to networks, and the program is designed to function silently: "This is undetectable right now so they can't see your ip address or log you." For example, the file advises, "Check out their desktop directory, there is always good stuff in there"; or "For personal info, names, telephone, addresses, family members names look in the My Documents directory." In response, Microsoft would be absolutely correct in stating that win-hack.c does not exploit any bugs or holes in the Windows operating system. It does, however, force one to reconsider the "feature" that allows anyone on the Internet access to your files and personal information.

Such a program that exploits one of Microsoft's "features" depends on a number of things in order to function. First and foremost, the feature has to be designed without regard to security. Rather than preventing users from doing something that puts their data at risk, the features tend to be set to lead users to leave themselves open to such risks by default. Second, users must have limited knowledge of how the computer works (file-sharing in this case), so that they don't know that they are putting their data at risk whenever they log on to the Internet. Finally, there has to be a sizable gap in expertise between those who exploit the feature and those who are exploited. These programs exploit very basic security flaws that are easily preventable if certain features are turned off or passwords are protected. As a result, those exploited tend to be users with the least knowledge and the least facility to protect their data.

While these basic exploits operate in the space between experts and end-users, there are two other programs that expose much greater problems with Microsoft's approach to security and that demonstrate the antagonism between hackers and Microsoft. These are the Cult of the Dead Cow's "Back Orifice" and the L0pht's program "L0phtCrack."

The 1990s were a time when hacking moved away from individual practice toward notions of group identity and political action. In the 1970s and 1980s, hackers had limited political agendas, and most of their actions were directed against one industry in particular, the phone company. More recently, in the wake of the AT&T break up, with the rise of the Internet, and with the increasing globalization of technology, hackers have begun to engage in more concerted political action, at both local and global levels. The results have manifested themselves in hacker groups engaging in political intervention, the formation of hacker collectives focused on enhancing hardware and software security, and the emergence of annual social events and hacker conventions.

From *Sub Rosa* to Sin City

In the early 1990s hackers found a way to emerge from the underground. As computers became an increasingly important part of everyday life, hackers gained increasing currency in the public and popular imagination. In the 1980s and early 1990s there were several tried and true methods for a hacker to make his or her name. First, hackers gained a certain prominence by virtue of their affiliations. To be a member of an elite circle of hackers, such as LOD or MOD, ensured a certain credibility in hacking circles. It also provided hackers with access to resources and information that were not more generally available. Within those small circles, hackers would learn from each other and generally develop their skills. Those circles also provided a network of information whereby hackers would learn the latest hacks or exploits.

A second and related means by which hackers made their reputations was by sharing information in public forums, either by disseminating text files throughout the underground or by publishing in underground journals such as *2600* or *Phrack*. These files, which usually consisted of basic material or information that was obsolete, demonstrated a basic mastery of systems or techniques. By the time these files reached the larger underground community, their information was at the very best dated and usually of little value. But the circulation of the information did document a hacker's prior successes. These files served as a means to solidify hackers' reputations, illustrating the degree to which they understood the systems they infiltrated. The means of distribution for these files was the electronic bulletin board system (BBS), which was usually run by a group of hackers or carried a group affiliation. An early issue of *Phrack* (June 1986) catalogued hacker groups and their affiliations. A typical entry would include a brief description, a listing of affiliated boards, and a listing of group members. The write-up for the Legion of Doom hackers read as follows:

LOD/H: Legion Of Doom/Hackers
These two groups are very closely intertwined. They both were formed on Plovernet. The founding member was Lex Luthor. Through the years, there have been LOD/H bulletin boards

such as Blottoland, LOD, FOD, and so on. Today there is Catch 22 and a new LOD BBS, supposedly being run by King Blotto. The current member list of the group is as follows:

Legion Of Hackers	Legion Of Doom
Blue Archer	Phucked Agent 04
Gary Seven	Compu-Phreak
Kerrang Khan	
Lex Luthor	
Master Of Impact	
Silver Spy (Sysop of Catch 22)	
The Marauder	
The Videosmith	

LOD/H is known for being one of the oldest and most knowledgeable of all groups. In the past they have written many extensive g-philes about various topics.[10]

Such a description would give other hackers several crucial pieces of information. It provides a bit of history, indicating the origins of the group and the group's founder or founding members; it identifies them with particular BBS systems or networks; and it provides a listing of current members, making it more difficult for hackers to pose as members of the group as a means to increase their credibility. Making it into such a listing gave hackers and hacker groups a kind of public certification and legitimacy and, because *Phrack* was being distributed nationwide, solidified hackers' reputations on a national scale.

Third, and finally, hackers can make a name for themselves (for good or ill) by getting "busted." Nothing legitimates a hacker's reputation more quickly than having law enforcement take an interest in her or his activities. Most commonly, hackers who are arrested face derision at the hands of their fellow hackers, for not being savvy enough, for being careless, or generally, for making stupid decisions that led to their apprehension. There is no doubt, however, that this is the primary means by which hackers gain public attention. It is also the basis for nearly every published, journalistic account of hackers and hacking. Hackers profiled in books, articles, and news stories are almost always hackers who have been arrested and prosecuted for crimes. *Phrack* also did its part in promoting the reputations of arrested hackers. The Pro-Phile feature was designed to enshrine hackers who had "retired" as the elder statesmen of the underground. The Pro-Philes became a kind of nostalgic romanticizing of hacker culture, akin to the write-up one expects in a high school yearbook, replete with "Favorite Things" and "Most Memorable Experiences." The "Phrack Pro-Phile," the editors write, "is created to bring info to you, the users, about old or highly important/controversial people."[11]

While the first two strategies are useful for gaining credibility in the hacker underground, the latter was usually the only means for a hacker to gain credibility outside the world of hackerdom. The 1990s would bring a transformation in the public nature of hacker culture. Starting in the late 1980s, hackers began to gather informally in an effort to meet one another face-to-face and share information. They would smoke, drink, and hack into the wee hours of the morning in the rooms of some unsuspecting hotel. The gatherings, initially, would be small, a couple dozen hackers at most at gatherings named SummerCon, PumpCon, HoHoCon (held over Christmas), and, perhaps most famously, DefCon and HOPE (Hackers on Planet Earth). One of the earliest, and most controversial, meetings was SummerCon 1987, which included, among other things, a number of arrests. The gathering was small and included informal discussion of hacking, BBS systems, and a range of other topics. As reported later:

> The full guest list of SummerCon '87 includes:
> Bill From RNOC / Bit Master / Cheap Shades / Control C / Dan The Operator / Data Line / Doom Prophet / Forest Ranger / Knight Lightning / Lex Luthor / LOKI / Lucifer 666 / Ninja NYC / Phantom Phreaker / Sir Francis Drake / Synthetic Slug / Taran King / The Disk Jockey / The Leftist / Tuc[12]

While fewer than twenty hackers were in attendance in 1987, that number would grow to nearly fifteen hundred a decade later.

Conventions provided the first organized efforts to bring hackers face-to-face in large numbers. While the conventions of the late

1980s would range from a handful to a few dozen hackers, the conventions of the 1990s would become full-scale events. The most widely known and recognized hacker Con is held in Las Vegas over the summer. Organized by Jeff Moss (Dark Tangent), DefCon is a three-day meeting that brings together hackers, security experts, law enforcement, and industry specialists to hear lectures and engage in discussions about computer security. The attendees are predominantly males, teens to early twenties, who are recognized by their handles rather than their names.

Part of what makes DefCon unique is that it openly invites industry and law enforcement to the gathering. There are even good-natured games, such as "Spot the Fed," where conference goers are invited to identify someone they think is a federal agent or law enforcement personnel and bring them up to the stage. The hacker states his reasons for thinking the person is "the fed," and the audience votes. If a general consensus is reached (or the suspected individual "confesses"), the hacker receives an "I spotted the Fed" T-shirt, and the fed receives an "I am the Fed" T-shirt. The contest is held between each speaker, and there is generally no shortage of willing participants on either side.

While speakers talk on issues ranging from how to hack the Las Vegas gaming industry to how to con your way into first-class travel, there are a range of "games," including a hacker scavenger hunt, electronic "capture the flag," where hackers take over one another's systems, and Hacker Jeopardy, hosted by *InfoWar* author Winn Schwartau.

Conventions are seen increasingly as places to share knowledge, meet "elite" hackers, and buy the latest hacker T-shirts. But they are also serving the purpose of organizing hackers in a way that had previously been impossible. The days' events and lectures are filled with messages about law, privacy, surveillance, and multinational corporations' dominance. Hackers are growing to see themselves as politically motivated out of necessity. A significant motivation has been the growth and dominance of Microsoft, a corporation that has been under the skin of hackers since Gates's initial confrontation with hackers over pirated software in the 1970s. In the 1990s, hackers began to respond not only by hacking Microsoft's software but by growing increasingly political in their agenda in response to Microsoft's policies.

Hacking Microsoft: cDc and Back Orifice

One of the oldest hacker groups in the computer underground calls itself the Cult of the Dead Cow (cDc). In addition to publishing an underground online journal, composed of cultural criticism and commentary, fiction, and hacking-related issues, the cDc members are dedicated to making themselves highly visible both in the computer underground and in mainstream culture. Unlike earlier hacker groups (such as LOD), who may have claimed that hacking was about learning, the media-ready Cult of the Dead Cow has a different approach, which is reflected in its mottoes: "Global domination through media saturation" and "cDc. It's all about style, jackass."

The cDc represents a major shift in both the computer underground and the media and mainstream representation of it. On the one hand, the nature of hacking has changed fundamentally in the last decade. With software such as Linux, it has become possible to completely dissociate hacking from criminality. On the other hand, media representations have come to focus almost exclusively on criminal aspects of hacking. As a result, as hacking has become less criminal in nature, representations of hackers have focused increasingly on that aspect of hacking. Hackers have become defined as "outlaws" not through their actions but through the process of representation. The cDc has utilized such representations to further its own agenda and has expanded the domain of hacking into the realm of the political, both locally and globally.

Oftentimes it is difficult to take claims of the cDc seriously, particularly when it issues its Global Domination Update or names Ebola as the "Disease of the Year." Such bravado tends to mask an important aspect of the subcultural identity that the hackers of cDc have formed. They are the first hacker group to take media representation seriously, and in doing so they are also the first group to work as critics of the mainstream incorporation of computer culture and incorporate the notion of political dissent into their history and identity. As one member describes the group's history:

Well.... There are those that would say that the Cult of the Dead Cow is simply the modern incarnation of an ancient gnostic order that dates back to the cult of Hathor, the cow goddess, in ancient Egypt.

Others may tell you that the Cult of the Dead Cow always has been, and always will be. A Universal Constant, if you will.

Of course, all these people are wrong.

In his book, "1984," George Orwell predicted a dystopia, peopled by soulless, spiritless, powerless drones, herded by a clique of absolute rulers, concerned only with maintaining their OWN POWER AT ALL COSTS....

1984.... Ronald Reagan is President, it is a "New Mourning in America."

In Texas, the heartland of America, the bastion of Patriotism and Old Time Religion, a small cabal of malcontents meet in secret.

They gather in a dark hovel, decorated with crude pornography, satanic iconography, heavy metal band posters and, most ominously, the skull of a DEAD COW....

As pirated copies of speedmetal and hardcore punk music play in the background, these malcontents speak of their disillusion with The American Way and their obsession with their new computers.

As the music plays, they form an unholy alliance, dedicated to the overthrow of all that is Good and Decent in America.

Realizing that a bunch of punk kids from Lubbock have as much chance of that as Madonna becoming Pope, they then decide to dedicate their lives to pissing off the establishment, becoming famous, and getting on TV.

Thus was born the Cult of The Dead Cow, scourge of the Computer Underground, Bete Noir of high school computer teachers worldwide, The Pivot of Evil for all who seek to blame the messenger, as well as their message.[13]

What makes the cDc's statement distinct from most hacker commentary is that it is positioned in terms of politics (for example, Reagan and "mourning in America"), and it uses a merger of discontent and technology to enact dissent ("these malcontents speak of their disillusion with The American Way and their obsession with their new computers"). It is also a basic theorizing of the idea of a technoscape in relation to the mass media. The cDc members realize that while they cannot control (or even influence) the mediascapes that impact their lives, they are in a position to attend to issues of technoscapes. That realization is one that connects technology to politics not in a metanarrative of control or change but in terms of a narrative of disruption.

The cDc represents, in that sense, a major break from the past in two ways. First, it is the first hacker group dedicated to a kind of political action based on principles of civil disobedience and visibility, and, second, it is the first group to connect hacker identity with the notion of political action. As Oxblood Ruffin describes it:

If there is one general theme that resonates with politics and hacking I would say that most people in the computer underground are varying shades of libertarian, but from my experience that doesn't really translate into group action. I know some people to be somewhat active politically, but I believe they act as individuals and not as part of a hacker group. I personally got involved with the cDc because I've been politically active, or working in political circles for a lot of my professional life. I saw the opportunity of using civil disobedience online — sort of another tool in the arsenal — but I don't believe that what we're doing is common, or even duplicated anywhere else.[14]

Their understanding of political action is not limited solely to underground activities. In fact, the goal of the cDc — unlike almost all other hacker groups, which did their best to function in secret — is to "become famous" and get on television. Of course, such descriptions are hyperbolic, but they reveal something about the transformation of the computer underground as well as the media representation of it. In the late 1990s, politics emerged as a central area of concern for the cDc as its members began to witness the globalization of technology and its political implications. One such instance has been the cDc's association with the Hong Kong Blondes, a group of dissidents in the People's Republic of China (PRC) who are utilizing hacking as a strategy for intervention. As Oxblood Ruffin described them:

The Hong Kong Blondes are a group of computer scientists and human rights activists who are committed to social change and democratic ideals in the People's Republic of China. They are especially interested in the relationship of the PRC to Hong Kong and are following the so-called *one china, two policies* doctrine quite closely. The Blondes are currently monitoring government networks and gathering data to be shared with other activists. . . . They would also be prepared to disrupt government/military networks in retaliation of any egregious human rights violation. . . . As to the risks involved, they are rather apparent: death, relocation, and loss of employment for family members, etc. The blondes are in this for the long haul and are hoping to contribute to the extremely difficult and very slow process of democratizing their country.[15]

The statement recognizes the power of both the globalization of technology and the globalization of resistance. In one incident, Lemon Li, a member of the Blondes, was arrested in China. As Blondie Wong, director of the Hong Kong Blondes, describes it: "Lemon was questioned in Beijing. She was released after a few hours but I couldn't take any chances so our associates moved her out of China. She is in Paris now. . . . She is acting more like traffic co-ordinator now. Much of our work is happening from the inside and she steers our efforts in the right direction."[16] Technological networking, the use of secure encryption techniques, and exploiting bugs and holes in network operating systems are allowing the Blondes to effectively communicate and coordinate political action around human rights violations.

While the cDc and the Hong Kong Blondes are working globally, cDc hackers have been working locally. On August 3, 1998, at DefCon, an annual meeting of hackers in Las Vegas, the Cult of the Dead Cow released a program called "Back Orifice," a play on Microsoft's NT software package "Back Office," which exposed major security flaws in Microsoft's Windows 95 and Windows 98 software. As the name indicates, the product was designed to rudely confront Microsoft and to force it to take notice of the program and the cDc itself. According to the security alert that the group released July 21, 1998, "Back Orifice is a self-contained, self-installing utility which allows the user to control and monitor computers running the Windows operating system over a network." The program, which runs transparently in the Window's environment background, has the ability to function as "an integrated packet sniffer, allowing easy monitoring of network traffic," and includes "an integrated keyboard monitor, allowing the easy logging of keystrokes to a log file." The implication is not only that data is vulnerable but that anything that is typed (including passwords) on the network is able to be viewed and stored by the Back Orifice program. Locations such as university computer labs or even a local Kinko's could easily have the program installed on a single computer, which would allow a hacker to remotely watch the network, monitor traffic for log-ins, and collect user-names and passwords from the entire network. In short, the very existence of this program makes any networked Windows 95 or Windows 98 machine suspect.

The release of the program was designed to focus attention on several serious security flaws in Microsoft's operating system. Microsoft denied that there was any security flaw in its product, claiming in a marketing bulletin, under the title "The Truth about 'Back Orifice,' " that the program "does not expose or exploit any security issue with the Windows platform. . . . In fact, remote control software is nothing new — a number of commercial programs are available that allow a computer to be remotely controlled for legitimate purposes, like enterprise help desk support."[17] The response reassured users that they were really at little risk and that as long as they followed "all of the normal precautions regarding safe computing" they would be fine. These suggestions included only downloading digitally signed and verified software and using a firewall (an intermediate machine between a computer and the network, or between a network and the Internet) to protect a network. Rather than offering a way to fix the problem or protect against unauthorized use, Microsoft simply denied the risk and tried to reassure customers.

The point that Microsoft failed to acknowledge was that Back Orifice was not designed to exploit bugs in the system but rather was intended to expose Microsoft's complete lack of concern about security issues. As one member of the Cult of the Dead Cow explained, "The holes that Back Orifice exposes aren't even really bugs, but more fundamental design flaws. Of course, Microsoft calls them Features."[18] As a result, "Back Orifice does not do anything that the Windows 95/98 operating system was not intended to do. It does not take advantage of any bugs in the operating system or use any undocumented or internal APIs. It uses documented calls built into Windows."[19]

In addition to allowing for complete remote control of another person's machine, remote keyboard logging, and network monitoring, Back Orifice has the ability to find and display cached passwords. These are passwords stored on the computer's hard drive, usually in an unencrypted form, to make the system more user-friendly. The kinds of passwords typically cached might include "passwords for web sites, dialup connections, network drives and printers, and the passwords of any other application that sends users' passwords to Windows so the user won't be inconvenienced by having to remember his passwords every time he uses his computer."[20] Again, it is essentially an exploitation of social relations, between the expert and the end-user, that makes such hacking possible. It is also the place where cDc takes aim at Microsoft. As one member of the cDc explains:

> Microsoft seeks to buffer the user from the actual workings of the computer. They give you a nice little gui [graphical user interface], integrated web-browser and all the bells and whistles. But why is there this file with all my passwords cached in plain text? Isn't that bad? Now-Now-Now, don't worry your head about that. Just watch the pretty pictures. Sleep... Sleep.
>
> The problem is that if Microsoft wants to buffer their customers from the workings of the computer, then they have to do a hell of a lot better of a job of protecting them from OTHER people who DO understand the workings of their computer.[21]

The issue is one of trade-offs. As Microsoft would later admit, "Windows 95 and Windows 98 offer security features tailored to match consumer computer use. This consumer design center balances security, ease of use and freedom of choice."[22] In essence, Microsoft recognizes that balance between convenience and security requires that some sacrifices need to be made, and those sacrifices can come at either end of the spectrum. A more secure system means that the user will have to understand more about how the system functions and may find it slightly more difficult to use (learning to password-protect shared drives, for example). A more convenient system, one that is easier to use, is likely to sacrifice security. Microsoft has chosen to err on the side of convenience, providing minimal security measures. In doing so, Microsoft has further widened the gap between hackers and end-users. What would appear to be a boon to hackers, easy access to the majority of computers being used both on the Internet and in business, actually presents hackers with a central problem. As hackers grow older and begin to find employment as computer programmers, security experts, and system administrators, they are faced with the task of making their own systems secure. In that sense, they inherit Microsoft's security problems.

For hackers, the problem, which is a result of the software, is being handled in the wrong place. The burden of security is placed on the end-user rather than on the software itself. As Microsoft claims, the "security features in Windows 95 and Windows 98 enable consumers to create a safe computing environment for themselves while preserving their freedom to choose which sites they visit and what software they download. However, neither operating system is designed to be resistant to all forms and intensities of attacks."[23] Rather than *being* a safe computing environment, the software *enables consumers to create* a safe computing environment. Again, rather than improving the software itself, Microsoft simply warns users to follow "reasonable and safe Internet computing practices, such as not installing software from unknown and untrusted sources," and adds a sales pitch for its more advanced operating system — "consumers whose computing needs require a higher level of security should consider Windows NT."[24]

The conflict is the same one that has been rehearsed ever since Bill Gates released his memo accusing the Homebrew Computer Club members of being thieves. Hackers see Gates and Microsoft as producing an inferior product, selling it for too much money, and taking advantage of a market of end-users that they were fundamental in creating. At a practical level, Microsoft is creating problems for hackers interested in securing their own networks, but at a philosophical level, Microsoft is violating the most basic tenets of computer culture. Most segments of computer culture, including the computer industry, have always been able to operate within the general confines of an ethic. This ethic was driven, for the most part, by the concept of a social conscience, a dedication to the principle that computers could make people's lives better. While that ethic has always been negotiated and even violated (early hackers taking Department of Defense money, Apple keeping its hardware proprietary, and so on), there has always been a genuine belief among hackers and industry that technology was doing more good than harm. That has also been a large part of the justification for hackers to remain politically neutral in all but the most local and immediate circumstances. Microsoft changed all that by embracing corporate policy that violated much of what hackers (and even industry) considered to be for the social good.

Most hackers and hacker groups would view those justifications as enough of a reason to release hacker programs, embarrass Microsoft, and force it to implement changes in its software. But for the cDc, the local and the global merge around the question of politics. Microsoft, hackers would argue, has become the high-tech corporation that they had always feared — multinational in scope and amoral in character. Companies such as Apple that had violated basic tenets of computer culture by keeping the hardware proprietary had also shown remarkable commitments to education, and Steve Wozniak, one of Apple's founders, had retired from Apple to teach at a local school. These kinds of balancing acts were often enough to keep hackers on the political sidelines. Whatever arguments one might have with Sun, Apple, or Intel, there was always something else to redeem them.

With Microsoft, the situation was different. Microsoft from the very beginning had operated in opposition to the ethic that animated hacker culture. When Microsoft challenged hackers on the grounds of their own ethics, however, a movement in the hacker underground was created that recognized politics as an essential part of a newly constituted hacker ethic. In response to Microsoft's challenges, which accused the cDc of being unethical for releasing its Back Orifice software, the cDc offered this response:

> We'd like to ask Microsoft, or more to the point, we'd like to ask Bill Gates why he stood shoulder to shoulder in 1996 with China's president and head of the Communist Party to denounce any discussion of China's human rights record at the annual meeting of the United Nations Commission on Human Rights in Geneva? Was the decision to cozy up to the world's largest totalitarian state based on some superior moral position, or was it just more convenient to trample human decency underfoot and go for even more money? Call us crazy, but we think that Microsoft has about as much right to condescend to the CULT OF THE DEAD COW as Li Peng does to lecture anyone who raises the issue of human rights abuses in China — a point of view that Bill Gates shares.[25]

As Microsoft opens up China as a new market with a billion potential customers, hackers argue that human rights has taken a backseat to profits. Such a trade-off, while not particularly shocking for most corporations, is something that hackers felt a computer company should take into account. Blondie Wong makes the connection explicit when he argues, "In 1996 he [Bill Gates] publicly endorsed China's position that human rights in China should not even be discussed at an annual meeting of the United Nations Commission on Human Rights. By taking the side of profit over conscience, business has set our struggle back so far that they have become our oppressors too."[26] Microsoft had, in the eyes of hackers, gone from being a tool in the struggle for freedom to part of the mechanism of oppression. The philosophy behind hackers' strategy of political intervention is based on focusing on U.S.-Chinese trade relations and American businesses, like Microsoft, trading with China. As Wong sees it:

> One of the reasons that human rights in China are not further ahead is because they have been de-linked from American trade policy. What this means is that when human rights considerations were associated with doing business with the United

States, at least there was the threat of losing trade relations, of some form of punishment. Now this just doesn't exist. Beijing successfully went around Congress and straight to American business, so in effect, businessmen started dictating foreign policy. There are huge lobbies in Washington that only spend money to ensure that no one interferes with this agenda. It's very well organized, and it doesn't end there.[27]

The intervention Wong suggests is "exposing them. By naming them and possibly worse," suggesting that hackers ought to use their "skills" to make business difficult for American companies trading with China. Indeed, the strategy of hacking as intervention is designed to make things "messy" for American businesses. But it is also a strategy for empowerment of youth culture:

Human rights is an international issue, so I don't have a problem with businesses that profit from our suffering paying part of the bill. Perhaps then they will see the wisdom of putting some conditions on trade. But I think, more importantly, many young people will become involved in something important on their own terms. I have faith in idealism and youth. It took us a long way in 1989. I believe that it will help us again.[28]

While visibility is an issue for the cDc, as self-described "media whores," it functions differently for members of the Hong Kong Blondes. When asked why he gave the interview, Blondie Wong responded:

Not for the kind of publicity you might think. We just need to have people know that we exist for now. It is like an insurance policy you could say. If anyone [of the Hong Kong Blondes] were arrested the possibility of execution or long imprisonment is quite real. In China, so much happens quietly, or behind closed doors. If someone is known, sometimes just that is enough to keep them alive, or give hope. So for that reason I'm saying we exist, that [we are doing] certain things. . . . It is not for fame, no. So this insurance policy, it is something that no one wants to use, but sometimes it is good to take precautions. This is my first and last interview. Now I can go back to being invisible.[29]

Being visible is viewed both as insurance and as risk. It is the problem hackers have faced since the 1970s, but with much higher stakes and to much greater ends. The focus on political action is almost entirely new. While hackers have traditionally viewed hacking as divorced from politics, the cDc and their involvement with the Hong Kong Blondes point to a new kind of engagement around questions of the globalization of technology. In large part, Microsoft's lack of attention to the basic tenets of computer culture and even the old-school hacker ethic have forced hackers to become more politically aware and politically active. The old-school ethic, which allowed for (and even encouraged) a multitude of sins, was entirely absent from the Microsoft philosophy. Microsoft had become the multinational corporation of Gibson's cyberpunk dystopia or the threat profiled in the film *Hackers*.

Hacking Microsoft II: The L0pht and L0phtcrack

While the cDc exposed flaws in the Windows operating system, a group of Boston-based hackers known as the L0pht had released a program designed to exploit security flaws in Windows's more advanced software, Windows NT.

The L0pht exemplifies a second trend in the hacker underground, the creation of confederations of hackers who come together to create, build, maintain, and break into their own systems. The idea behind a group such as the L0pht is to create a safe space for hackers to experiment without risk. The L0pht has gained its reputation in the underground not by breaking into other systems, defacing Web pages, or hacking government sites, but by finding holes in software and writing security releases that explain how the flaws make certain systems vulnerable. The L0pht members are the leaders of a new trend in hacking that positions hackers themselves as the watchdogs of the computer industry.

The L0pht is a group of roughly half a dozen hackers who meet outside of their regular jobs in a Boston loft to explore issues of hardware and software hacking. Their projects have ranged from hacking the MacIntosh computer to translating old software to run on handheld PCs such as the Palm Pilot. What they are best known for, however, is finding and exploiting holes in UNIX-based and Microsoft operating systems and software applications.

Mudge, one of the L0pht's best-known hackers, is also responsible for one of the best-known, and most often exploited, bugs in the UNIX environment. The bug, called a "buffer overflow," exploits a certain feature in the way that UNIX executes programs. Certain software is written to handle information in terms of a "buffer." Buffers set expectations for the computer's memory. For example, if I create a variable called "name" and expect it to always be less than 256 letters long, I might set the buffer for the variable "name" to 256. If someone enters a name longer than 256 characters, the program gets an error and stops running; the program crashes. What is interesting to the hacker mind-set is what happens when the program crashes. As Mudge discovered, the program will attempt to execute whatever code is left over. If you have written your program carefully, then, you can simply fill up that buffer with two things: garbage and a program. By filling up the buffer with garbage until it overloads, you make the program crash. When the program crashes, it executes whatever code is left over. If you calculated correctly, what is left over is your program. So to run a program, you simply fill up the buffer with garbage, put the code on top of it, and crash the program. There are some programs that a common user is not allowed to run because they present security risks. The advantage of the buffer overflow is that the hacker can use it to run programs not normally accessible to the average user. Because the system (rather than the user) is running the program, the hacker can use such a trick to take control of the system. For instance, when a program needs to access system resources that generally aren't available to the average user (for example, hidden password information), the program automatically changes to a "super-user" status, giving the program access to everything on the machine. If the "buffer overflow" targets that program, it also runs the hacker's code as a "super-user," giving him or her unlimited access to the machine.

Such programs are based a fusion of two discoveries, both of which illustrate the manner in which transparent understanding allows hackers to exploit systems that most users don't understand or have access to. First, using a buffer overflow (or any other number of tricks or hacks) requires a basic and intimate knowledge of how the machine works and of how the operating system processes commands and executes code. But that alone is not enough. These hacks also require an understanding of how the machine copes with errors, what the computer does when things go wrong, when someone accidentally or purposefully "smashes the system." From there, it is a matter of finding out how to make the system fail — by doing what the computer sees as impossible or by providing irregular or excessive input. In short, hackers simply do what is unexpected or extreme in order to watch the results. If the computer is designed to handle a 256-character name, hackers feed it a 257-character name, a 500-character name, or a 10,000-character name and observe what happens. For hackers, these machines are at their most interesting when they fail, when they break down, or when they respond to situations or data they are not meant to handle. To the hacker, the most important thing is that watching machines cope with errors is a way to understand them better and more fully. It is the ultimate moment of transparency, discovering for the first time how the machine reacts to a certain bit of code or piece of information.

In keeping with the tradition, the exploits that the L0pht publicizes about UNIX usually lead to patches, bug fixes, and alerts in the tried-and-true method. In contrast, the L0pht's confrontations with Microsoft have been somewhat less cordial. In 1997, members of the L0pht discovered a series of holes in Windows NT, Microsoft's "New Technologies" line of software that was being widely disseminated and marketed as a secure network server system. The hole that the L0pht discovered allowed hackers to remotely query the Windows server, ask it for a list of passwords, download those passwords, and then crack them. To make matters worse, Microsoft had made Windows NT backward-compatible with its old DOS-based software called LANMAN. The older, less secure LANMAN system of passwords was over a decade old and produced passwords that were easy to crack with very simple algorithms that had been widely circulated for years. When the NT server was validating a password,

it would ask the machine making the request which form to use —
the new NT password system or the older, less secure LANMAN
system. If LANMAN was chosen, it would convert the new, more
secure NT password into the less secure LANMAN version.

One of the primary differences in passwords was the length and
case of the text that was used. Windows NT allowed for passwords
up to sixteen characters in length, including both upper- and lowercase letters, but LANMAN had only allowed for eight characters, all
in one case. A password such as "ThisIsMyPassword" (an acceptable
NT password) would be truncated and converted to "THISISMY"
for the LANMAN system. With a basic desktop computer, it is possible to run through a list of one-case, eight-letter passwords in a
matter of a few days. A list of passwords sixteen characters long,
both upper- and lowercase, in contrast, can't be broken in a reasonable amount of time. Again, Microsoft's desire to have machines
that were "backward-compatible" with 1980s technology resulted
in a system that was much less secure.

When the bug was discovered, the hackers informed Microsoft,
which responded by denying that the problem was serious. The
official press release from Microsoft read as follows:

> Use of the L0phtcrack tool requires getting access to the
> Administrator account and password. This is not a security
> flaw with Windows NT, but highlights the importance of
> protecting Administrator accounts and reinforces the impor-
> tance of following basic security guidelines. If customers follow
> proper security policies, there is no known workaround to get
> unauthorized Administrator access.[30]

Sensing a potential public relations problem, Microsoft took the position that if system administrators were careful there would be no
problem, an approach that would come back to haunt the corporation. Because the L0pht hackers had programmed the cracking
program in a UNIX environment, and because it required a complex command structure and was seen as difficult to master, the
press wasn't terribly interested either. So a few hackers, computer
geniuses, can break into a server? We already know they can do that.
The hackers at the L0pht took a page from Microsoft's PR book. If

the bug wasn't enough to get either Microsoft's or the media's attention, they would have to find something that would. A few weeks
later, the L0pht had rewritten the program for the Windows operating system, given it a graphical user interface, and released it, free
of charge, on the Web. Now anyone who had the time to download
the software and read a short help file could go about cracking NT
servers anywhere on the Net. The program, L0phtCrack, got the
attention of everyone, including Microsoft, which sent out press releases initially denying the problem and later admitting it once it had
created a patch for the hole. The fixes Microsoft came up with were
easy to get around because they tended to hide, rather than solve,
the problem. As a result, the L0pht hackers have been continually
improving their software to keep up with Microsoft's changes. With
each iteration, Microsoft hides the problem and the L0pht finds it
again. This type of "security through obscurity" strategy is never
very effective against hackers, primarily because they already know
all the best hiding places. Obscurity, while very effective against a
neophyte, is hardly a challenge for a hacker who knows the inner
workings of the machine. When the L0pht released its later version of
the software, L0phtcrack 2.0, Microsoft again responded similarly:

> L0phtcrack 2.0 does not expose any new issues in Windows
> NT or Windows 95. L0phtcrack 2.0 does not enable non-
> administrators to directly access passwords or password hashes
> on Windows 95 or Windows NT systems.[31]

This statement is *technically* true. However, information about
passwords is obtained whenever requests to log in are made, so
passwords and password hashes are *indirectly* accessible. As the documentation to L0phtCrack explains, all you need to do is ask for a
log on and the computer will send you its password information. A
program called a "sniffer" will do just that — request log on and
password information and store it for later hacking. As a result, the
hacker does not need access to the machine, to an administrator
account (as Microsoft had claimed), or even to a user account. All
hacking can be done remotely, without even gaining access to the
machine.[32]

What is remarkable about a program such as L0phtCrack is that it targets Microsoft so directly and yet has none of the "criminal" or "underground" dimensions that usually characterize hacker software. It is a serious piece of software that security professionals might purchase to check out their own systems. And, indeed, the L0pht is marketing it that way. The reason that Microsoft attracts these hackers' attention and inspires their wrath is precisely that the corporation seems to have little or no concern for security. It ignores warnings, avoids making fixes until absolutely forced into it, and winds up supplying inadequate fixes when it does release them. In sum, Microsoft relies on complexity and secrecy to maintain any level of security for its systems, and when its secrets are exposed, Microsoft simply denies that they were secrets at all.

The hackers of the L0pht see themselves as providing a public service. They are truly concerned about security and believe that the only way a company such as Microsoft will ever make its products secure is to be shamed into it. As Mudge explained in response to an article in *Mass High Tech,* which accused hackers and the L0pht of being "crackers who inflict chaos":

> I was completely specific on what type of "chaos," as you put it, the L0pht "inflicts" (thank you for two words with negative connotations). The exact same type that consumer reports does. To wit: if I am using a piece of software and find it to be flawed — we go public with it. This alerts the general populace to the problem and forces the company to fix it. So...out of this chaos you, as an end-user, see technological and security related enhancements. Sorry if that is so evil.[33]

The members of the L0pht see themselves as educators about issues of security, fulfilling the same function as *Consumer Reports,* and they see themselves doing this in a similar manner. While, as a group, the L0pht may be doing important security work, they are also very much engaged in other aspects of hacker culture, projects ranging from creating secure wireless communications to finding new and better ways to make pornography run on Palm Pilot, handheld computers.

Hackers such as those at the L0pht are working essentially to expose what amounts to corporate secrets, hidden from public view.

But the notion of corporate secrecy has also taken on a particular sense of importance around the issue of law and regulation. While hackers such as Mudge are busy breaking corporate codes of secrecy, the law is finding it increasingly important to protect those secrets in an environment that is making such a task increasingly difficult. As a result, the L0pht members, along with a number of other hackers, have begun to recognize that political engagement is a vital part of their own hacker ethic. Over the past several years, hackers from the L0pht have testified before Congress, have routinely monitored and written responses to House and Senate bills about privacy, security, and information regulation and policy, and have continued to release advisories and product updates about computer software that may contain bugs or flaws that make systems vulnerable to outside attack.

While hackers have grown increasingly political in response to the current climate of the corporatization and globalization of technology, the political climate in regard to hackers has shifted as well. In the late 1990s, computers and computer security have taken on a heightened importance in almost every element of life — from ATM machines to grocery store checkouts. In these contexts, the figure of the hacker, while becoming increasingly organized and political, has been continually framed as an outsider, as a threat, and as a danger.

Part II

Hacking Representation

.

Hacking Representation

This part explores hacker culture through an examination of hackers' relationship to mainstream contemporary culture. In particular, what such analysis reveals is that hackers actively constitute themselves as a subculture through the performance of technology and that representations of hackers in the media, law, and popular culture tell us more about contemporary cultural attitudes about and anxiety over technology than they do about the culture of hackers or the activity of hacking. Although these representations provide an insight into contemporary concerns about technology, they serve to conceal a more sophisticated subculture formed by hackers themselves.

In particular, I explore hacker culture through a reading of the online journal *Phrack* (the title is a neologism combining the terms *phreak* and *hack*), a journal written by and for the computer underground, and by examining the ways in which hackers have responded to and occasionally embraced cultural representations of hacker culture.

One of the principal factors that makes hacking possible is the contemporary culture of secrecy that governs a significant portion of social, cultural, and particularly economic interaction. This culture of secrecy has produced a climate in which contemporary hackers feel both alienated and advantaged. Although hackers philosophically oppose secrecy, they also self-consciously exploit it as their modus operandi, further complicating their ambivalent status in relation to technology and contemporary culture. This part explores the themes of secrecy and anxiety in relation to both contemporary attitudes toward technology and the manner in which hackers negotiate their own subculture and identity in the face of such cultural mores.

Chapter 4

Representing Hacker Culture: Reading *Phrack*

Hackers have always relied on communities to share ideas, information, and access to technology. In the 1950s and 1960s those communities were based on physical proximity and common interest, particularly in the computer labs of MIT, Harvard, Cornell, and a handful of other universities. What these early hackers learned they shared with each other, trading code and even access to computers. In such a context, it was easy to determine who was doing the most interesting and innovative work. The problems that were being solved, however, were mostly technical, questions of how one makes a computer do a particular task, perform an operation, or solve a problem. The earliest hackers were using local technology to solve local problems.

By the 1960s, the split between the hackers who stayed attuned to the local problems that they solved in their labs and the hackers who had found ways to integrate their love of technology with the larger social and political movements of the 1960s had become marked. The latter, who produced *TAP* (*Technological Assistance Program*), began a tradition of producing literature and disseminating information that was not only technical, but also practical. The literature of the hacker underground spread word about the latest hacks, whether it be about rerouting long-distance telephone calls for free, hooking up your own power connection, or understanding the political agendas of the phone company. The literature had the primary goal of "getting the word out," of transforming local solutions into national, and even global, ones.

In the early literature of *TAP* and other underground newsletters, there was a basic recognition that information is power and that

115

secrecy, on the part of institutions, governments, and corporations, was a means by which power was maintained and consolidated. Telling people how to do simple things to make their lives better was at the root of the underground literature, one that spawned what Bruce Sterling has called an "anarchy of convenience," a system of "ingenious, vaguely politicized rip offs."[1] These low-level scams, which would eventually outlast the political movement that produced them, were what fueled the spread of the Yippie message. More fundamental to the Yippie ideology was the idea that technology, particularly free use of phones, provided a centering mechanism for the movement as a whole, a technological infrastructure that members of the movement could access and usurp as their own.

In 1983, "Tom Edison," then editor of *TAP*, had his computer stolen and his apartment set on fire, and the journal ceased publication. Two other hacker underground periodicals sprang up to fill the gap left by *TAP*'s disappearance. In 1983, Eric Corley, under the pseudonym Emmanuel Goldstein, started *2600*, a three- or four-page printed pamphlet that talked about the tyranny of the phone company and the political responsibility people had to be aware of its power. *2600: The Hacker Quarterly*, in the spirit of former hacker publications, also provided information about how one might best the phone company and what tips and tricks might be used to avoid its watchful eye (and its expensive charges). Even the name, *2600*, was an indication of the new journal's intent. As noted earlier, *2600* referred to the tone, 2600 Hz, that was used by blue boxes to seize control of long-distance phone lines and allow phone phreaks to make free calls.

Alongside Corley's *2600* was *Phrack*, a conglomeration of the terms "phreak" and "hack," which was published exclusively in electronic form. Directed at the new generation of computer-literate hackers, *Phrack* was a collection of files that contained technical information but that also served as a guide to the underground, a who's who of computer hacking. Where *2600* would take up the political side of hacking, *Phrack* would emphasize the culture of hacking. These journals, both of which are still published, reveal a great deal about the new school of hacking, both in terms of its political and social agendas and in terms of its culture and awareness of representation. Reading these documents as representative of hacker culture, however, requires a basic introduction to the ways in which hackers think about the literature they produce.

Reading Hacker Culture

Hacker culture, which appears at first glance to be easily accessible and widely represented, poses some curious difficulties for a cultural reading. In large part, the rhetoric concerning hackers and hacking is hyperbolic. Therefore, any reading that aspires to understand hacking as a cultural phenomenon must both account for and move beyond the exaggerated dimensions that represent and misrepresent the culture. Such a reading would be fairly straightforward if hackers were not prone to precisely the same kind of overstatement and mischaracterization of their activities that the media and government officials are. It is not uncommon for hackers to boast of their abilities to "crash the phone system" or how easy it might be for "one person to take down the majority of the Internet," even as they claim that they never would do such a thing. A second and related problem with examining hacker culture is the manner in which it deploys secrecy. Hackers present themselves as "elite," meaning not only that they have access to certain information and knowledge or possess certain abilities but also that they are members of a kind of loosely affiliated "inner circle" to which only the smartest, most gifted hackers belong. Ironically, the clearest indication of one's "elite status" is visibility. In essence, the more "elite" one becomes, the better known one is. As a result, the high-status, private club of "elite hackers" exists only in relation to its public visibility. Hence, it is not surprising to find that much of the discussion of hackers and hacking has grown up around crises, particularly arrests and legal battles.

Hacker culture has had a sizable and significant online presence since the early 1980s. That culture developed around systems of the electronic bulletin board (BBS). The BBS was a computer system, usually run from a hacker's home, that would function as a kind of community center or public meeting area. Although initially, and for a long time, a BBS could only accommodate one user at a time, there were message databases (similar to a bulletin board) where hackers could post messages and communicate with one another. The BBS was the perfect medium for an underground culture to develop and flourish. As Bruce Sterling writes: "Computer bulletin boards are unregulated enterprises. Running a board is a rough-and-ready, catch-as-catch-can proposition. Basically anybody with a computer, modem, software, and a phone line can start a board. With second-hand equipment and public-domain free software, the price of a board might be quite small — less than it would take to publish a magazine or even a decent pamphlet."[2] Hackers could leave messages about new discoveries, new hacking methods, or personal matters. Before the Internet (up to the mid-1990s), the BBS was the primary means of communication among computer hackers. Bulletin boards would often have primitive e-mail systems allowing users to send private messages to each other, and, perhaps most important, they would have file areas that would make accessible computer programs and information files that would be useful for hacking.

Cultures emerged around different boards and even produced some of the first hacker groups. Hackers would gain status and reputation by writing files describing how to perform certain hacks and by writing programs that would aid others in their hacking. Some of the first programs to surface were "WarGames Dialers," which would call every phone number in an assigned prefix and record which numbers responded with modem tones, similar to the program Matthew Broderick's character employs in *WarGames*. These numbers would then be checked later for possible hacking exploits. The BBS also formed the basis for the first hacker groups. When a hacker would write a file and post it to a BBS, he or she would often sign it, not just with his or her pseudonym but also with his or her home BBS, for example, Taran King of Metal Shop AE. Files and messages would be passed from BBS to BBS, occasionally making it from one coast to the other. The system was informal but did serve to associate hackers with particular boards, areas, or hacker groups. Such systems were also ideal for passing along illicit information. Because of the difficulty in regulating them (or even knowing about their existence) and their ability to offer complete anonymity to their users, the boards provided the ideal "safe space" for hackers to congregate, share information, learn from one another, and build their reputations.

The tradition of passing files from one system to another was the central means of disseminating information about hacking, as well as about hacker culture. Files that explained the technical aspects of hacking would also discuss the cultural, ideological, and even political aspects of it. These collections of files would be the genesis of the first underground hacker publications. What *2600* had put down on paper, *Phrack* was doing online.

Since the early 1980s, *2600* has been dedicated to the dissemination of "forbidden knowledge," focusing particularly on the knowledge of how things work and function. One of the premises that guides Goldstein's thinking is that it is much more difficult for institutions and governments to do bad things when they are held to public scrutiny. Cynicism also underwrites this thinking. Governments and institutions will continue to do bad things *unless* their secrets are exposed and they are confronted by their own misdeeds.

2600 has been the champion of hackers who have been arrested, convicted, and imprisoned and has done a great deal to bring public attention to such cases. *2600,* while published in New York, runs in each issue a list of meetings that take place all over the country. These meetings are a way for hackers to find one another, to share and trade information, and to generally make connections in the hacker underground. The meetings are often scheduled in "in-your-face" locations. For example, the New York meeting takes place in the lobby of the Citicorp Center, "near the payphones." Should a hacker be unable to attend, the meeting list usually provides the number of the nearby payphones so he or she might call in. *2600*'s message is not just being spread in the United States. Nearly half of the listings are for international meetings, including those in Argentina, India, Russia, and South Africa, in addition to most major European countries.

In many ways, *2600* has taken up the mantle of the 1970s and 1980s but has continued it in 1980s and 1990s style, adding publications to the Internet, holding two hacker conferences (H.O.P.E. [Hackers on Planet Earth] and H2K [H.O.P.E 2000]), running Internet mailing lists, hosting a weekly public access radio show ("Off the Hook"), and continually working to foster new generations of hackers. The content of *2600* has always come from contributors (rather than a staff of writers). Each issue is introduced with a message from the editor about issues of political or social importance and is usually filled with letters from readers either asking questions or commenting on past articles.

2600 has always reflected a particular social and political agenda for hackers, one that has followed the old-school ethic that "information wants to be free." What it reveals about hacker culture is the premium that is placed on knowledge, and particularly difference. For Goldstein, there is a definite sense of a "hacker perspective," described in *2600* as "what happens if you do this instead of what everyone else on the planet does."[3] Knowledge, to Goldstein, is about freedom, but it is also about power. Knowledge and its dissemination are the means to monitor and hold responsible (and more than occasionally embarrass) those in power.

In many ways, *2600* reveals a great deal about the history and commitments of hackers and hacking. Its political and social message, however, tells us relatively little about the underground world of hackers as a culture. *2600* upheld the tradition of hacker publications. Two years after its beginning, teenage hackers in the Midwest created something uniquely of the 1980s that would, for the first time, create a full-blown sense of hacker culture.

Reading *Phrack*

In 1985, two hackers going by the handles Knight Lightning and Taran King began putting together a resource for the computer underground that appeared on their BBS, "The Metal Shop AE." Those resources were collected into *Phrack,* which would publish informational articles of interest and use to hackers and phone phreaks. Three things make *Phrack* unusual as a part of underground hacker culture and different from periodicals such as *TAP* and *2600*. First and foremost, with minor (and a few major) interruptions, *Phrack* has continued in basically the same form, spanning more than fifteen years. Electronic journals have no material infrastructure — no phones, no buildings, no presses, no trucks to deliver them, no stores to sell them. As a result, they tend to spring up, publish a few issues, and vanish. Unlike most electronic underground journals that disappear routinely, *Phrack* has had staying power. It has even survived the changing of hands through several editors and a high-profile court case. Second, even though it is one of the most popular and widely read hacker journals, it has always remained free and accessible in electronic form. Indeed, it has never moved from its original mode of distribution, a collection of files electronically distributed. Third, *Phrack* has always provided a dual function. While one of its primary purposes is to disseminate information about hacking and phreaking, it also serves as a cultural focus for the hacker community — it tells the members of the underground what is going on, who has been arrested, who is angry with whom, and so on. Unlike its predecessors, *TAP* and *2600, Phrack* is designed to organize what had previously been a loose confederation of people, files, ideas, and gossip that had found their way onto various BBSes or computer files.

It can be said that *Phrack* has had its finger on the pulse of hacker culture. Three main features compose *Phrack*'s cultural side: the brief or alternatively rambling introductions to each issue that explain, more often than not, what is going on with the editors and staff of *Phrack,* most often with a justification why the issue is later than expected; the *Phrack* Pro-Philes, which provide biographies of selected hackers, detailing what got them interested in hacking, as well as their various explorations, exploits, and affiliations; and, perhaps most important, the "Phrack World News," which details the rumors, gossip, and news of the hacker underground.

The two initial editors of *Phrack,* Knight Lightning (Craig Neidorf) and Taran King (Randy King), did little of the technical writing for the journal, instead calling on the general hacker community for contributions. Like *2600,* many of the initial articles were less technological in focus and were more about "forbidden knowledge." Articles ranging from "How to Pick Master Locks," by Gin Fizz and Ninja NYC, to "How to Make an Acetylene Balloon Bomb," by The Clashmaster, were the focal points of the first issue of *Phrack.* The informational content would evolve, but it was significantly less important than the other cultural information that the journal provided. In fact, the majority of the information that *Phrack* published about hacking and phreaking was widely disseminated already, either as independent philes or in other technical online journals. *Phrack*'s real news was about the culture of the computer and hacker underground. Sterling describes the editors of *Phrack* as "two of the foremost *journalists* native to the underground."[4] In part, as Sterling speculates, the emergence of *Phrack* in the Midwest during the mid-1980s was to compensate, at least in part, for the fact that St. Louis (the location of *Phrack*'s home BBS) was hardly the center of the hacking world. New York and Los Angeles were the primary centers for hacking. But *Phrack* was in touch with both the New York and Los Angeles scenes and was reporting essentially on each of them to each other. In a very short period, *Phrack* became the standard publication for the underground scene. If you were to run a *serious* underground board, including *Phrack* in your collection of files for download was mandatory: "Possession of *Phrack* on one's board was prima facie evidence of a bad attitude. *Phrack* was seemingly everywhere, aiding, abetting, and spreading the underground ethos."[5]

Phrack was also essential to spreading the reputations of the various groups. Periodically, *Phrack* would list or profile active groups, and "Phrack World News" would always document the comings and goings of new and old groups. That *Phrack,* like most hacking periodicals, was always more about the cultural aspects of hacking than about its technical aspects is evidenced by its increasing focus on providing social updates to the hacking community. Even the technical aspects that *Phrack* published were more about constructing and building the *ethos* of the author than about providing cutting-edge information. In fact, the central goal of *Phrack*'s technical side seemed to be a consolidation and dissemination of information already known to the broader hacking community. What *Phrack* did differently was fly in the face of the culture of secrecy.

Unlike *TAP* or *2600,* which were both print publications, *Phrack* faced a series of new problems. As an electronic journal, freely distributed, *Phrack* faced challenges from the governments, institutions, and corporations it frequently took as its targets.

Reading the Culture of Secrecy: E911 and Copyrights

If *Phrack* is known at all outside of hacking/phreak circles it is, iron-ically, because government and law enforcement personnel read it to keep tabs on the hacker underground. That relationship between the underground and law enforcement has also led to two very interest-ing developments regarding the underground journal itself. The first is a somewhat bizarre prosecution of *Phrack* editor Craig Neidorf for publishing a supposedly proprietary and potentially damaging Bell-South document, the second is *Phrack*'s equally bizarre copyright agreement. In both cases, what is at stake is the question of how one comes to "own" information and how that question of owner-ship impacts directly on the creation and maintenance of a culture of secrecy.

In 1988, a hacker who went by the handle "The Prophet" entered a BellSouth computer system and downloaded a file that documented improvements in the 911 emergency services that were soon to be implemented. The 911 document read, in part, as follows:

WARNING: NOT FOR USE OF DISCLOSURE OUTSIDE BELLSOUTH OR ANY OF ITS SUBSIDIARIES EXCEPT UNDER WRITTEN AGREEMENT.[6]

The document, referred to as E911 (E for enhanced), and its pub-lication in *Phrack,* became the center of a controversy at 2:25 P.M. on January 15, 1990, the day that AT&T's computers caused a mas-sive telephone service failure across the United States. The history of these events, their interrelationships, and the details of their after-math have been documented in Sterling's *The Hacker Crackdown* as well as Paul Mungo and Bryan Clough's *Approaching Zero,* among other works.[7] The Prophet's hack was documented in *Phrack* in two parts — the first was the publication of the BellSouth document, "Control Office Administration of Enhanced 911 Service"; the sec-ond was a follow-up document published with the self-explanatory title "Glossary Terminology for Enhanced 911 Service." Both doc-uments were published by The Prophet under the pseudonym "The Eavesdropper," out of fear that using his normal handle could lead to his arrest.

BellSouth responded with characteristic appeals to fear. As Mungo and Clough report, "In the hands of the wrong people, BellSouth technicians said, the critical E911 document could be used as a blueprint for widespread disruption in the emergency systems. Clearly, hackers were the wrong sort of people. According to Bell-South, 'any damage to that very sensitive system could result in a dangerous breakdown in police, fire, and ambulance services.' "[8] By the time the document appeared in *Phrack,* it had been edited to less than half its initial size, and "phone numbers and some of the touchier and more specific information" had been excised.[9] According to the hackers involved, the purpose of publishing the document was never to disrupt phone or emergency services but, instead, to "reflect glory on the prowess of the underground and embarrass the telcos."[10]

The incident may well have passed unnoticed if it wasn't for the AT&T phone crash of 1990. A year after the E911 document had been published, in the wake of the AT&T crash, Knight Lightning (*Phrack* editor Craig Neidorf) was interrogated by federal agents, local authorities, and telephone company security officers. Neidorf was being accused of causing the crash of AT&T's system. The visit occurred two days *after* AT&T had announced that a glitch in its own software had caused the crash. Agents, apparently unconvinced by AT&T's own explanation, were convinced that the computer underground was involved or, perhaps, could be involved if hackers chose to be. The AT&T crash gave law enforcement the incentive to go after Neidorf, *Phrack,* and the computer underground as a whole. As Sterling argues, "the consensus among telco security — already horrified by the skill of the BellSouth intruder — was that the digital underground was out of hand. LOD (Legion of Doom) and *Phrack* must go."[11] The resulting investigation would lead to an indictment against Neidorf that would have him facing ten felony counts resulting in up to sixty-five years in prison.

The prosecution's case rested on two arguments. First, it argued that the E911 document was *dangerous* — that in the wrong hands it could do a tremendous amount of damage to what was portrayed as a very fragile emergency-services operation. Second, the document (twelve pages in length) was incredibly valuable to BellSouth, and theft of it from the corporation's computers constituted a major crime. The document was valued, according to BellSouth officials, at $79,449. How the figure was arrived at was not known until well after the trial, but as John Perry Barlow remarked, one can "imagine an appraisal team of Franz Kafka, Joseph Heller, and Thomas Pynchon."[12] There were two problems with the prosecution's case. First, the information, which was considered so dangerous and proprietary, was already available as a BellSouth publication to anyone who took the time to call in on the corporation's 800 number and request it. And, second, the document that contained everything that *Phrack* had printed and a whole lot more cost thirteen dollars.

Neidorf was cleared of all charges but was left with legal bills of more than one hundred thousand dollars. The E911 document would emerge again later, this time in the trial of Robert Riggs (The Prophet) and two others. This time, the E911 document would be valued at $24,639.05. (As Mungo and Clough point out, "the 5 cents [was] presumably included to indicate that the figure had been very accurately determined.")[13] The value of the "stolen" E911 document was of crucial concern to law enforcement and prosecution teams.

The initial $79,449.00 value was calculated by Kim Megahee, working for BellSouth. To arrive at the final figure, she simply added up all the costs of producing the document, including $7,000 for writing the document, $6,200 for a project manager to oversee the writing of the document, $721 for typing it, costs for editing, mailing labels, indexing, and so on. Of course, those were only the basic costs; there were also "hardware costs," including $850 for a computer monitor, $31,000 for the VAXstation II that the document was written on, $22,000 for a copy of "Interleaf" software, and $2,500 for the VAX's VMS operating system.[14] The number would later be reduced to reflect research-and-development costs, reducing the total to the $24,639.05, which would be the final amount used to prosecute Riggs.

The importance of the question of the document's value can not be emphasized enough, but the answer to that question follows the wrong course if we pose it first in monetary terms. What is really at stake in the value of the E911 document? Two things, both of which point to a single conclusion: the value of the document is the value of the secret, since BellSouth was never deprived of the information, only the exclusive right to that information. First, BellSouth was deprived of the secret of the E911 document itself. When the document was publicly and widely distributed, it lost status as a secret. Second, and perhaps more important, BellSouth, having its document copied and distributed, was itself exposed as not able to maintain secrecy, which is the real source of damage. BellSouth, in having the document copied, lost nothing other than the exclusive right to control how that information was disseminated and how it was made public.

The culture of secrecy that surrounds technology can not be separated from the question of the ownership of information. The secret is not about withholding information; quite the contrary, the culture of secrecy is about limiting access and to whom information is given.

In the 1990s, *Phrack* would fight a second battle involving secrecy and ownership, this time using the notion of ownership of information to limit access and preserve its own secrets. In order to do so, *Phrack* editor Erik Bloodaxe (Chris Goggans) issued the following copyright statement in *Phrack:*

> Nothing may be reproduced in whole or in part without written permission of the Editor-In-Chief. Phrack Magazine is made available quarterly to the amateur computer hobbyist free of charge. Any corporate, government, legal, or otherwise commercial usage or possession (electronic or otherwise) is strictly prohibited without prior registration, and is in violation of applicable US Copyright laws.[15]

Phrack's copyright agreement is structured to expose, violate, and mock precisely the culture of secrecy that it enacts by its very presence. It says, in essence, we will distribute *Phrack* freely to anyone who does not participate in the culture of secrecy that we seek to overthrow, namely, the corporate, government, legal, and commercial interests. *Phrack* is, by the account of one of its former editors, Chris Goggans, designed to be an affront to precisely those interests that are prohibited from getting the journal free of charge. Goggans describes the first issue he took over as editor: "I think that one went over very well. That issue was pretty hilarious because I had a lot of stuff about packet switching networks, and it was a big slap in the face to B. T. Tymnet. I had a whole lot of fun with that issue."¹⁶ The point of *Phrack* for Goggans is to violate the secrecy that preserves the *proprietary* structure of Tymnet. The value of the information *Phrack* provides for the hacker community is only partly about use. The greatest value attached to the information is the idea that *Phrack* is telling big corporations' secrets. The "slap in the face," as Goggans describes it, is the informational value of *Phrack*. It is the violation of secrecy that makes *Phrack* a valuable source of information, not merely the information itself.

The use of the copyright, and the subsequent effects it produced, should be viewed as more that a simple game of tit for tat. What is revealed in the institutional structures of privacy and secrecy tells us a great deal about society's relationship to technology and the culture of secrecy that is so ensconced within that relationship.¹⁷

The copyright is nothing more than the mobilization of an entire institutional structure that is designed to continually reinforce and reinstitutionalize secrecy. A copyright does more than simply demarcate authorship: it specifies the means, manner, and ability to disseminate the information. It is the very condition of the secret. Secrecy never operates as an absolute principle. In order to be a secret, information must, at some level, be shared. There must be those who "know" in order for the secret to function. If information is not shared, secrecy loses its force, even its meaning. The true power of the secret is in its disclosure, and in that sense, like technology, secrecy is *relational* rather than technical. The copyright agreement in *Phrack* gains its force not from what is communicated in *Phrack* but because of who receives it, because of the relationships it fosters, generates, sustains, and, in some cases, even makes possible:

Corporate/Institutional/Government: If you are a business, institution or government agency, or otherwise employed by, contracted to or providing any consultation relating to computers, telecommunications or security of any kind to such an entity, this information pertains to you. You are instructed to read this agreement and comply with its terms and immediately destroy any copies of this publication existing in your possession (electronic or otherwise) until such a time as you have fulfilled your registration requirements. A form to request registration agreements is provided at the end of this file. Cost is $100.00 US per user for subscription registration. Cost of multi-user licenses will be negotiated on a site-by-site basis.[18]

The institutional structure that ensures secrecy (of which copyright is only a single example) is revealed in this single gesture — copyrighting a publication. What is revealing is what that gesture makes both possible and impossible. In short, it specifies the condition of ownership of information; or, if we think of it in a slightly different register, it names a certain set of relations now attached to that information (of which authorship/ownership is only one). In essence, the mark of copyright elevates any information that bears it to the status of a secret. As regards *Phrack,* this is certainly the case — assigning *Phrack* a copyright, essentially, allows it to remain a secret from the government, law enforcement, or corporate interests that might seek it out. Nowhere is this more clearly demonstrated than in the plea that those who do subscribe will only do so under the condition of secrecy itself. ("We're going to pay, but don't tell anybody we're going to pay.")

The coup de grâce, for *Phrack* is represented in a final reversal. Because *Phrack* had only one paying subscriber, Goggans knew that those who should be registering were still receiving the magazine. In doing so, they were themselves retreating into another layer of secrecy. That is, they were receiving the magazine in secret. At this point the roles have been completely reversed — the hacker, now employing the institutions of secrecy, is himself being subverted by those who seek to undermine those very principles. But in *Phrack*'s case there is an all-important difference — *Phrack* doesn't really care about the profit; it is only interested in challenging or disrupting the basic power relationship. In fact, for perhaps the very first time, corporate and hacker interests are aligned. The point (and irony) was not lost on Goggans, who immediately responded by calling the corporations on it: "I named several people who were not only getting the magazine but in one case, they were spreading it around and, of course, none of them even contacted me for registration. I had a riot with it. It was a lot of fun."[19]

As a journal that stands in violation of the very principle of secrecy and the proprietary nature of information, it seems extraordinarily odd that *Phrack* would choose to copyright its information. But that oddity begins to make a great deal of sense upon closer examination. The idea of the copyright is deployed against those who utilize it for protection of the secrecy that *Phrack* seeks to dismantle. As Goggans tells the story, "After I took it [*Phrack*] over, I went ahead and registered it with the Library of Congress and I filled a DBA as Phrack magazine and for the first issue I put out a license agreement, sort of, at the beginning saying that any corporate, government, or law enforcement use or possession of this magazine without prior registration with me was a violation of the Copyright Law, blah, blah, blah, this and that and Phrack was free to qualified subscribers; however, in order to qualify as a qualified subscriber, one must be an amateur hobbyist with no ties to such a thing."[20] In short, Goggans deployed the institution of secrecy against those who have constructed it. *Phrack* was able, at least in principle, to operate in secrecy (or at least have knowledge of those persons who were receiving *Phrack* and who might have government, corporate, or law enforcement interests) by claiming the same sort of ownership of information that had been used against it in the case of the E911 document several years earlier. If *Phrack* was to be watched or monitored, this agreement was designed to make sure that those who ran *Phrack* could monitor the monitors.

The effect that *Phrack*'s copyright produced was perhaps even more interesting and revealing than the copyright itself. The copyright never prohibited anyone from getting *Phrack* but only charged a fee of one hundred dollars for a subscription to government, law enforcement, and corporate organizations. According to Goggans, the copyright "went over like a ton of bricks with some people. A lot of corporate people immediately sent back, 'Please remove my name from the list.'" But perhaps the most interesting responses came from corporate people who, in an effort to negotiate the problems of secrecy, responded by saying, "We're going to pay, but don't tell anybody we're going to pay." With one exception, no one ever did, and Goggans, in typical *Phrack* style, utilized that as fodder for the next issue of the journal, "saying that all of them are lying, cheating scums and that they have no respect for our information so why should they think it odd that we don't have any respect for theirs."[21]

Hackers' relationships to secrecy, then, have to be understood at two different levels — first, secrecy is that which must be destroyed on a global level, which is to say on a corporate, governmental, and administrative level; but, second, it must be deployed strategically on a local level. For the hacker, secrecy is capable of eliminating access to information, but it is also capable of providing access to information within a culture that values and enacts secrecy itself. Accordingly, while certain things must be disseminated publicly, other things must be kept secret. The tension is not an easily resolvable one, and *Phrack* documents how this tension manifests itself throughout the underground. There are two particular, recurring "features" in *Phrack* that illustrate the complexity of hackers' relationships to secrecy, to technology, and, perhaps most important, to each other. Those features are the *Phrack* Pro-Philes, short biographies of well-known hackers, and the "Phrack World News," a reading of local, national, and international news from a hacker perspective.

Reading *Phrack* Pro-Philes

Even as secrecy is one of the central notions that defines the computer underground, hackers can be seen as having an ambiguous relationship to it. In order to hack, or at least to hack successfully, hackers must enact the very secrecy that they are working to dismantle. Specifically, hackers code themselves through the use of handles, pseudonyms, affiliations, and group or club memberships. Known by these electronic pen names, hackers gain prominence by demonstrating their knowledge of computer coding, phone systems, social-engineering skills, or other valuable hacking abilities. To be respectable as a hacker, one must have a command of readily available resources — one must have the ability to program in C, proficiency in UNIX shell navigation, and so on; to be "elite," one must be able to break the culture of secrecy that surrounds technology.

Once a hacker has the ability to demonstrate these proficiencies, he or she must find a way to attach his or her pseudonym to those particular proficiencies. For example, a hacker particularly adept at programming may write and release a program that makes routine a set of complex or laborious hacker tasks (such as the repeated dialing of phone numbers when searching for modems in a particular phone exchange). A hacker who understands packet-switched networks may write an article or phile describing how these networks operate. If the information is considered valuable enough, it might come to be included in an issue of *Phrack* and thereby be disseminated on boards throughout the country.

The articles that generate the most attention, from hackers as well as computer security experts, are the ones that go a step beyond the dissemination or consolidation of already available information. These articles, such as the E911 document, serve as *evidence* that the hacker is able to go one step farther — he or she can tell you something that isn't available elsewhere, something that someone doesn't want you to know. That is, the elite hacker is capable of violating the culture of secrecy that defines the current state of technology. What defines the value of the information is not necessarily how *useful* the information is but, rather, how *secret* the information is. Some of the most potent hacks garner completely *useless* information from a practical standpoint, but they are invaluable in exposing precisely what it is that technology has kept hidden.

In this way, a doubled sense of the idea of *knowledge* emerges. Knowledge of technology means both *knowing how technology works* (in a practical sense) and *knowing what technology hides*. This doubled sense of knowing is most fully realized by the hacker not as a binary construction but, rather, as a supplementary one — technology works *because* technology hides. As a result, the technological always stands in relationship to the hidden or secret, and it is that relationship that is always open to exposure and exploitation.

Hackers who understand the relationship between technology and secrecy possess not just knowledge but a kind of knowledge that sets them apart. But that setting apart, in relation to knowledge, is about authority and authorship. As a technology itself, authorship participates in the dual sense of knowledge. It both performs a practical function and hides something. The *Phrack* Pro-Phile is one of the ways in which these dynamics of knowledge are played out in detail.

To be featured in a *Phrack* Pro-Phile, one must have attained a certain stature in the hacker community. The feature was introduced as follows: "Phrack Pro-Phile was created to bring information to you, the community, about retired hackers or highly important/ controversial people. In this issue, we bring you . . . "[22] The structure of the Pro-Phile mimics the layout of an FBI "wanted" poster — listing both name and alias, date of birth, height, weight, eye color, hair color, and so forth. Of course, there are some important differences. Where a "wanted" poster begins with "Name:" followed by "Aka(s):" (or "Aliases:"), the Pro-Phile begins with (and privileges) the alias or handle and treats the name in a rather dismissive manner, replacing the word "Name" with the instruction "Call Him." In doing so, the Pro-Phile both inverts the structure and reprioritizes it — the handle functions as the name; the real name functions as a familiar kind of greeting. The Pro-Phile follows with a list of information that focuses on physical description; this mirrors the "wanted" poster, with the exception of occasional inclusions of extremely personal (and obviously false) information, such as "sperm count," "blood type," or "shoe size." In doing so, the Pro-Phile becomes a parody of an FBI profile, suggesting that hackers have an awareness of both what is revealed in and concealed by law enforcement's perspective on them. The personal data section of the Pro-Phile mimics and critiques law enforcement's efforts at description, illustrating the manner in which law enforcement violates and enacts secrecy in its pursuit of hackers.

The second section of the Pro-Phile is essentially an open interview, where the hacker pro-philed responds to a series of questions. First, the hacker describes his or her "Origins in the Phreak/Hack World" and "Phreak/Hack BBSes." From there, the interviews differ, but almost all include listings such as "Knowledge Attributed To," "Interests," "Phreak/Hack Groups," "Favorite Things," "Most Memorable Experience," "People to Mention," "Inside Jokes," and, finally, a "Serious Section." The Pro-Philes serve three main cultural functions. First, they provide and locate a sense of history for a diverse and disjointed (and self-described anarchic) community. Second, they function normatively to help define what constitutes acceptable and unacceptable hacking behavior, both in terms of the activity (for example, what constitutes malicious hacking) and culturally (for example, what sort of music hackers enjoy). Third, the Pro-Philes construct and elevate hackers as cultural icons or heroes.

Of particular interest throughout the Pro-Philes are moments of origin, which function as narrative accounts of the rise (and oftentimes the fall) of the hacker. In almost every case, *Phrack* Pro-Philes begin with the heading, "Origins in the Phreak/Hack World," a section that documents that particular hacker's relationship not necessarily to hacking but to technology itself. For example, one of the earliest hackers to be pro-philed, "Karl Marx," describes his entrée to hacking as such: "Manufacturing Explosives — He wanted to blow up his High School."[23] More often, though, the narrative unfolds as one of exploration and discovery: "A friend of Bloodaxe's father bought a MicroModem II to get information from Dialog for his legal practice. He still remembers the first time he used it. His friend's dad used Dialog through Telenet. Once he saw Telenet, he began trying various addresses. One of the first things he ever did was get into a 212 VAX/VMS with GUEST/GUEST."[24]

The introductions to these Pro-Philes are littered with references to the first generation of personal home computers — Atari 400s, Apple IIs, TRS-80s, Amiga 1000s, Commodore 64s, and so on. In that sense, the history of hacking is tied directly to the birth of popularly available PCs. But there is a second category that shares equally in the birth of hacking, the "Origins in Phreak/Hack BBSes." While the technology was important, almost every one of the pro-philed hackers had an interest beyond the technology itself — their interest was in networks. With few exceptions, these hackers ran a BBS (bulletin board system) out of their homes, and all of them participated on various pirate software or hacker-related BBSes. The history of hacking is tied to the practical aspects of technology (computers and networks), but it also has its root in the culture of secrecy. Most narratives detailing hackers' early exploits involve them piercing through a veil of secrecy — usually by means of a happy accident or via an introduction — to enter a world that was known to only a few. For instance, while the technology was an important first step, Erik Bloodaxe primarily used his earliest computer and modem for trading pirated software. When Erik and a friend went to trade a new game to another Atari user, "a guy named Devious Xevious traded them something called Software Blue Box for it, and gave them a BBS to call: Pirate-80. In 1983, Erik Bloodaxe entered the hack/phreak world. He was blue boxing [calling for free] most of his calls by then."[25] The world into which hackers enter is defined not by physical space but rather by its separation from the mainstream. It is a world about secrets that operates in secret. Its origins are located both in the technology itself (as a practical dimension) and in what technology makes it possible to hide, namely, BBSes.

After establishing the origins of hacking, Pro-Philes provide a fair amount of information about the hackers themselves. Although they often contain exaggerations, inside jokes, and a fair amount of bragging, they also clearly define the pro-philed hackers as experts capable of speaking as cultural icons. Throughout each piece, the pro-philed hacker gives advice: "Hacking. You can read all the gfiles in the world, but unless you actually go out and hack, you're going to remain a novice. Getting in systems snowballs. It may take you a while to get in that first one, but after that it becomes easier and easier."[26] Or, alternatively, Karl Marx's advice on gaining experience: spend "long hours pouring over Bell Systems Tech Journals from 1970–present. He suggests to anyone who wants to learn non-trivial, but useful things — or who just wants to get some really *powerful* vocabulary for social engineering — try using your local college or large public library."[27] Such advice-giving also constructs the hacker as a kind of "elder statesman," one who is experienced, worldly, and knowledgeable.

The *Phrack* Pro-Philes, finally, are also an effort to construct a sense of community around a set of authors who constitute the basis for that community. At that level, it is not surprising to find that the majority of the Pro-Philes focus on "retired" hackers. No longer active, these hackers (most in their early to mid-twenties) pass into a state of veneration or even apotheosis. Almost all of the hackers pro-philed have also been arrested, signaling a kind of symbolic death, which makes it difficult or impossible for them to continue hacking. It is also a culturally constructed narrative of male sacrifice for the community. Accordingly, *Phrack* Pro-Philes constitute a certain kind of "death" of the hacker (or the hacker's handle) that is culturally celebrated. Such transformations not only are inevitable but are, in fact, part of the structure of authorship itself — the "age-old conception of Greek narrative or epic, which was designated to guarantee the immortality of the hero. The hero accepted an early death because his life, consecrated and magnified by death, passed into immortality; and the narrative redeemed his acceptance of death."[28] What is at stake for the hacker is the very question of authorship.

In current speculation about the state of authorship, there is a repeated refrain that the author is dead or has disappeared. Foucault calls for us to use this critical moment and to reexamine the empty space left by the author's disappearance; we should attentively "observe, along its gaps and fault lines, its new demarcations and the reapportionment of this void; we should await the fluid functions released by this disappearance."[29] Nowhere could Foucault's call be more apt than with the discourse on hackers and writing. The hacker, after all, *must disappear* in order to hack, and yet must *not disappear* in order to be a hacker.

The hacker, through the use of a handle, calls attention to the act of authorship, announcing that she or he both is and is not who he or she claims to be. Quite self-consciously, then, the name of the hacker, like the name of an author, "remains at the contours of texts — separating one from the other, defining their form, and characterizing their mode of existence. It points to the existence of certain groups of discourse and refers to the status of this discourse within a society and culture."[30] What this troping on authorship allows is for the hacker to both be known and yet remain anonymous. What the *Phrack* Pro-Philes allow for is the reemergence of the author after the fact. Once "retired" or out of the scene, the hacker's handle is both sacrificed and immortalized as the author reemerges, no longer a hacker, but now an author. No longer is he or she anonymous; instead, she or he reemerges to claim the acts done under another name. The technology of authorship, for the hacker, serves the practical function of naming and defining discrete

textual activity (hacking itself, technical articles on other information) and maintains the structure of the secret in that it separates
the hacker from his or her proper name and physical description.
The appearance of the hacker signals the disappearance of the subject. The appearance of the Pro-Phile signals the reemergence of the
author and the death of the hacker.

In the symbolic death of retirement or "getting busted," there is
a reversal. The connection of the author to the proper name and
physical description signals the *disappearance* of the hacker and
transforms the hacker yet again. What appears in the space opened
up by the exposure of the secret cancels out the practical aspect of
the hacker. With the secret exposed, the hacker disappears.

Thus, in essence, the function of the *Phrack* Pro-Phile is to make
the hacker *disappear* and replace him or her with a hero — one who
is to be remembered through the narrative of his or her untimely
death and who returns through the structure of narrative. Where
the Pro-Phile serves the function of celebrating individuals, a second
feature of *Phrack,* the "Phrack World News," gives hackers a filtered
reading of current events that allows them to position themselves
within a broader narrative of hacker culture.

Documenting the Underground: "Phrack World News"

What initially started as simply "News" would later evolve into
"Phrack World News," an accounting of the goings on in the hacker
community from the most mundane and trivial, including personality wars and group feuds, to mainstream news items that affected
hackers. For instance, in "News II," the second compilation of news
items by Knight Lightning, we find this juxtaposition:

> MCI/IBM Merge. MCI Telecommunications company has
> merged with IBM and their phone industry SBS. This was
> an effort to join the two as strong allies against AT&T. IBM
> computers vs. AT&T computers. MCI Telecommunications vs.
> AT&T Telecommunications. Changes arising from this merger
> (if any) are not known, but none are expected for some years.

Followed by

> Overlord 313 Busted: Step-dad turns him in. Overlord's step-dad always would be checking his computer to see what was on it and what was nearby. Last week he noticed the credits in Overlord's file on Wiretapping, which can be seen in this issue of Phrack. He reported his findings to Overlord's mom. She had a talk with him and he promised to stop his evil ways. His step-dad didn't believe him for a second.[31]

Ironically, at this point in *Phrack*'s history, the second story was probably the more important of the two. In later years, particularly as hackers began to grab national headlines, the nature of "Phrack World News" would change. The only standard for reporting is whether or not the news is of interest to hackers and the hacker community. Important information includes what BBSes are up or down, who has been arrested, what new groups have formed, which have disbanded, and what hackers have chosen to retire.

"Phrack World News" serves as a filter that doesn't distinguish mainstream news from events of hacker culture, and oftentimes information, reports, or news stories are reframed, titled, or retitled by hackers in order to make a particular point. In one instance, a copy of the San Diego police department's "Investigators' Follow-Up Report" titled "Damage Assessment of and Intelligence Gathering on Illegal Entry (Hacking) Computer Systems and the Illegal Use of Credit Cards" was included in "Phrack World News" under the title "Multiplexor and the Crypt Keeper Spill Guts." The details of the investigation and arrest of the two hackers are given in several sections, including follow-up reports from investigators, e-mails, and an "Aftermath" section. The story is followed by a letter sent from Kevin Marcus, titled "The Crypt Keeper Responds," in which he provides a detailed explanation of the events surrounding his arrest. The letter ends with an explanation as to why he is not a "nark" and a mea culpa:

> If I were a nark, then I would probably have given him a lot more information, wouldn't you think?
> I sure do.

I am not asking anyone to forget about it. I know that I screwed up, but there is not a whole bunch about it that I can do right now.

When Sadler was here asking me questions, it didn't pop into my mind that I should tell him to wait and then go and call my attorney, and then a few minutes later come back and tell him whatever my lawyer said. I was scared.[32]

As *Phrack* evolved in the 1990s, so did its news section, ultimately resulting in a kind of hacker news clipping service, where *Phrack* regulars could read reprints of articles about them, about hacking generally, or in which they were quoted or featured. In the news section, one is likely to find a story from the *New York Times* juxtaposed with a letter from a hacker or a reprint of a news story from an obscure journal or news service. As hackers began to make the mainstream news, *Phrack* continued to disrupt the authority of traditional forms of news, usually by demonstrating how coverage of hackers was one-sided, failed to tell the whole story, or left out essential details. What began as an informational resource for hackers would later become essential PR for the underground, focusing primarily on the relationship between hackers and law enforcement. While *Phrack* would feature mainstream news articles, it would also offer hackers a chance to respond to articles or features about them or to tell "their side of the story" in response to what had been written about them. In doing so, "Phrack World News" both demonstrated the manner in which hackers were covered (by reprinting the articles) and provided a corrective measure by filling in important pieces of information that revealed biases or hype in such coverage.

As a result, "Phrack World News" was able to recode current news stories in a way that made recontextualization possible and also left the hacker readership more informed than they would have been had they read a short news blurb in a local paper. It was, in every way, an attempt to control the news for the select readership of *Phrack*. That control carried with it two important connotations: first, that the reports one would read in "Phrack World News" were the stories that were being disseminated to mainstream culture; and, second, that those reports were always and inherently flawed.

"Phrack World News" was both inauthentic (because it was not the news that the majority of the population would see) and *more* authentic (because it told both sides of the story in much greater detail). As a result, "Phrack World News" provided both a detailed accounting of events and an accounting of how those events were being reported.

As hackers increasingly made their way into the mainstream media, "Phrack World News" changed substantially, transforming from an informational resource in its earliest incarnation to a media watchdog in its later form. In both cases, however, "Phrack World News" has served an important role of reporting the news and re-contextualizing it in a way that hackers can use to better understand their position in mainstream culture and as a forum to respond to that position.

As hackers found themselves more frequently in the news, they also found the underground disappearing. And as *Phrack* became more widely available, its mission changed. In issue 56 (May 2000) Shockwave Rider in his Pro-Phile pronounced the death of the underground at the hands of the Internet:

> The underground is no longer underground. Forums which once existed for the discussion of hacking/phreaking, and the use of technology toward that end, now exist for bands of semi-skilled programmers and self-proclaimed security experts to yammer about their personal lives, which exist almost entirely on the awful medium known as IRC. The BBS, where the hack/phreak underground grew from, is long since dead. Any chump can buy access to the largest network in the world for $19.95 a month, then show up on IRC or some other equally lame forum, claiming to be a hacker because they read bugtraq and can run exploits (or even worse, because they can utilize denial-of-service attacks). The hacker mindset has become a nonexistent commodity in the new corporate and media-friendly "underground."[33]

The sheer volume of hacking news has made it impractical for *Phrack* to keep up-to-date with events as they happen and has spawned a secondary site, attrition.org, that specializes in monitoring hacker news, reporting on inaccuracies, and correcting errata in news reports. Additionally, hackers have taken to creating their own news sites such as the Hacker New Network and Anti-Online, both of which provide a specialized news service about events and news of interest to the hacker community.

Conclusions

Phrack has played an essential role in the creation and maintenance of the computer underground. It has survived a high-profile court case and has continually responded to changes both in the computer underground and in mainstream culture. *Phrack,* perhaps more than any other single vessel, has communicated the standards of the underground and has functioned to create an elite class of hackers who would gain prominence by spreading information or being the subject of Pro-Philes or news.

Phrack demonstrates that the computer underground's culture is a rich one, with heroes and villains, mythologies and lore, and a world-view that, while fundamentally at odds with that of mainstream culture, both colors and is colored by news and current events. In short, while *Phrack* does impart information to the hacker underground in its articles and exploits, its more important function has been in creating a culture for the underground and in transmitting news, gossip, and lore about the hacks and hackers that define hacker culture. In doing so, *Phrack* established itself as essential reading for the culture of the underground and as a result had a central and defining role in shaping what that culture would look like for nearly fifteen years.

For hackers, *Phrack* has provided a venue in which they could be known without facing the risks of being known. *Phrack* served as the means to legitimate hackers for the underground, both by presenting them as celebrated heroes to the readers that made up the underground and by simultaneously taunting a larger audience of government officials, institutions, and corporations by presenting forbidden information and exposing secrets.

Chapter 5

(Not) Hackers: Subculture, Style, and Media Incorporation

The computer underground emerged in large part in journals such as *Phrack* and *2600*. As I have argued throughout, however, it is impossible to separate the representations of hackers from the creation of hacker identity. As a subculture, hackers have developed a particular sense of style that has been transformed over time and has been structured as an increasingly fluid and powerful form of resistance. As a youth culture, hackers are continually looking for ways to perturb or disrupt authority and challenge any understanding or representation of who they are. In tracing out the manner in which hacker style has developed, mutated, and evolved, I examine below its beginnings in computer culture, its transformation as a locus of subcultural identity, and, ultimately, the moment of collision between forces of subcultural identity among hackers themselves and forces of media representation in the 1995 film *Hackers*.

Understanding hackers of today necessitates a basic understanding of the history of the subculture that preceded them and of how, traditionally, subcultures have functioned to resist dominant cultural interpretations of them.

Youth Culture Online

Hackers are not the only youth culture online. In fact, youth culture seems to have found the Internet to be the preferred medium for expression. Issues that have typically represented youth culture — rebellion, resistance, fan culture, music, fashion, and pop culture — all find expression in Internet chat rooms, World Wide Web pages, e-mail mailing lists, and assorted other online elements. For example, traditionally problematic aspects of youth and teen culture, such as lesbian and gay identity, are finding varied forms of expression on the Internet.

Hackers, however, illustrate a particular aspect of online culture that is more properly called a subculture, a culture that is both inherently tied to a larger, in this case, parental culture, but also resistant to it. Subcultures are marked by their fluidity, their constant shifting both in meaning and in the processes by which meaning is made. As Dick Hebdige describes it, the meaning of subculture is always in dispute, and style is the area in which opposing definitions clash with most dramatic force. This tension is the site where items of common interest or importance overlap, where a subcultural element seizes control over the meaning of an object that has importance for the larger culture. The meaning of style, then, is generated from the subcultures' appropriation of a symbol of mainstream culture. That appropriation also involves a reinterpretation. In this way, subcultures take a piece of the larger culture and recontextualize it in order to give it different, oftentimes oppositional, meaning. As Hebdige argues, it is more than a response — it is a dramatization that commonly takes a forbidden form in either language, expression, or action.[1] At this level, subcultures enact a style of bricolage, the means by which objects are rearranged and recontextualized to give them different meanings and construct new discourses. This style is also fundamentally disruptive to the larger social and cultural discourse of which the subculture is a part.

Subcultural style, then, is about identifying objects of cultural semiotic importance and repositioning them in oppositional ways both to signal a refusal of the mainstream discourse and to construct a new discourse around these repositioned and rearticulated objects. Not surprisingly, a primary location of subcultural identity is youth culture. Through music, fashion, literature, and graffiti, for example, we see the radical semiotic reconstruction of the world through the eyes of youths.

Subcultural identification is also about resistance to authority, and in particular it is a resistance to the methods, styles, and mannerisms of the larger, parental culture. For youth it is about the transition from a world of parental authority, where the parents dictate how things are to be done, to a world of responsibility, where youth make decisions for themselves. The transition is marked by rebellion, defiance, and a seemingly single-minded focus on difference.

The importance of online culture — particularly in the 1980s, 1990s, and up to the present — for youth culture is grounded in three factors. First, the youth of the 1980s and 1990s are the first generation to grow up more computer-literate (and generally technologically literate) than their parents. For that reason alone, technology represents a way of doing things, a style, that is radically different from that of their parents. Second, technology, and computer culture more specifically, is constantly in flux. Such a fluid environment not only allows for radical recontextualization but demands it. Computer culture and computer style are in many ways the ideal hotbed for youth rebellion, as they require constant change in keeping with hardware and software developments. Third, and perhaps most important, the semiotic space that technology presents is one that is considerably less material than the traditional outlets of expression. Fashion, music, and literature, three primary outlets of youth culture expression, require a primary material component that is able to be marked, transformed, or reappropriated by mainstream culture. Computer culture, in contrast, is much less material in nature. While the hardware, the actual technological component itself, is material, the software and the style (the means by which one does things) are not.

If we contrast something as fundamental to youth culture as music with the Internet, it becomes clear that the difference of expression rests on materiality. For music to be expressed, it needs a material venue. Whether that venue be a public performance, a nightclub, a recording studio, or a local gathering, that material element represents a point of intersection between subculture and parent culture. That point of intersection is one that allows for the parent culture to prohibit expression (for example, arresting a performer for indecent lyrics) or to recontextualize that expression itself (for example, the moment when rap music is integrated into advertising campaigns). These two cultural responses, which Hebdige has defined as incorporation, are the means by which mainstream and parent cultures recoup meaning.[2] The first, ideological, incorporation is the means by which culture reacts to a subculture either by transforming it into something that is prohibited or by trivializing it or domesticating it in such a way as to render it meaningless.

Alternatively, the threat can be neutralized in an altogether different way, by turning the difference that subcultures represent into a commodity, whereby that difference can be bought and sold, marketed, and exploited. At that moment, the power of subcultural difference is neutralized as it is fed back into the mainstream culture and marketed as difference. Once difference is turned into a commodity, its meaning becomes frozen and its subversive power is lost. With computer culture, beyond basic questions of access, there are almost no material constraints on subcultural coding and production of meaning. Web pages require no printing presses, chat rooms require no public meeting space, and the lack of physical appearance makes style a purely semiotic exercise. In short, by making subcultures virtual, online culture becomes fluid and increasingly resistant.

The Origins of Computer Style

Hacker culture, born in the computer labs of East Coast universities in the late 1950s and 1960s, was a result of computer programmers doing everything in their power to beg, borrow, or steal computing resources. As a result, these hackers would often be forced to find time in the late hours of the night and into early morning, using less than ideal machines, and oftentimes being forced to work out clever compromises or work-around solutions to accomplish the task before them. The solutions, or "hacks," permeated early computer culture, eventually becoming central to it. For example, hackers at major computer labs would engage in a process of "bumming" code. Each hacker would contribute to a program by finding increasingly elegant solutions to programming problems. The focus on elegance was a measure of how sophisticated one could be in programming. The simpler the solution, the greater value it was seen to have. The goal of "bumming" was to reduce the number of lines or commands necessary to accomplish a certain task. The more clever or elegant the solution, the more cultural capital the hacker would accrue.

This way of thinking about problems, which necessitated thinking in nontraditional, often outrageous ways, was at odds with the
dominant thinking about computer programming. While the students were busily hacking late at night in the labs, their professors
were offering courses in "structured programming," a style that presumed that a single, superior, mathematically precise solution existed
for each problem encountered. The conflict is exemplified by an interchange between Richard Greenblatt, one of MIT's original hackers,
and Edsger W. Dijkstra, a mathematician and an original proponent
of structured programming. As Sherry Turkle describes the legendary
interchange:

> In Dijkstra's view, rigorous planning coupled with mathe-
> matical analysis should produce a computer program with
> mathematically guaranteed success. In this model, there is no
> room for bricolage. When Dijkstra gave a lecture at MIT in
> the late 1970s, he demonstrated his points by taking his audi-
> ence step by step through the development of a short program.
> Richard Greenblatt was in the audience, and the two men had
> an exchange that has entered into computer culture mythology.
> It was a classic confrontation between two opposing aesthetics.
> Greenblatt asked Dijkstra how he could apply his mathematical
> methods to something as complicated as a chess program. "I
> wouldn't write a chess program," Dijkstra replied, dismissing
> the issue.[3]

The problem, as it turns out, with Dijkstra's position is that people,
ultimately, *did* want chess programs and lots of other programs
as well, which made it necessary to think differently about programming. The difference, however, was not necessarily about
programming. The triumph of bricolage is, ultimately, an end-user
phenomenon. People prefer to play with computers, rather than
program them.

This dichotomy has been true since the first PC was mass-
marketed. I have described the Altair 8800, the first PC (ca. 1974),
a number of times above. Here I simply want to point to the way
it emphasized programming — not only did the machine have to be
assembled and soldered together by the hobbyist; it also came with

no software, which meant that the owner had to program the machine, initially by toggling switches on or off to produce a particular set of results in a series of lights on the front of the computer's case. The machine had no keyboard, no monitor, and no hard drive or floppy disk. It was also the site of the first major hacker/industry controversy.

In the mid-1970s, a computer whiz at Harvard dropped out, moved to New Mexico, and began writing software for the Altair 8800. His first program was a port of the computer language BASIC, made to run on the Altair. The computer whiz's name was Bill Gates, and the company that he founded to write that software was Microsoft. The hobbyists who owned these new Altair computers had begun to find each other and form loosely knit groups. They also engaged in the process of "bumming code," working to improve each other's programs and programming skills by revising code. When these hackers got a hold of Altair BASIC, they set about distributing the code and working to improve it. Rather than paying the twenty-five-dollar fee to Microsoft, these hobbyists would make copies for their friends and give them away. As noted earlier, one of the most famous hobbyist clubs, the Homebrew Computer Club in the San Francisco Bay area, gave away free copies with the stipulation that those who took a free copy should make two additional copies and give each of them away for free as well.

The style that hackers adopted, particularly in places such as Homebrew, thus relied on an extensive knowledge of computer programming and, as was the case with the Altair, computer hardware and engineering as well. Premiums were put on experimentation and a style of "play," which usually meant utilizing resources that were already at hand, rather than purchasing or adapting commercial applications. By the mid-1980s, the computer industry had begun to incorporate this style of play into the applications for commercial distribution. At that point, anyone could be a hacker simply by purchasing products made by AUTOCAD, Lotus, Microsoft, or VISICALC. As Turkle sees it:

The revaluation of bricolage in the culture of simulation includes a new emphasis on visualization and the development of intuition through the manipulation of virtual objects. Instead of having to follow a set of rules laid down in advance, computer users are encouraged to tinker in simulated microworlds. There, they learn about how things work by interacting with them. One can see evidence of this change in the way businesses do their financial planning, architects design buildings, and teenagers play with simulation games.[4]

I want to further Turkle's analysis by suggesting that the shift from a culture of bricolage in the production of computer software to the culture of bricolage (and what she calls "simulation") in the consumption of computer software represents an important moment of incorporation of hacker-subculture style.

At this moment, bricolage was transformed. What started as a blockage in the system of representation, a radical new way of doing things and thinking about computers and programming, had been reduced to a commodified form of style, mass-marketed for popular consumption. Bricolage, or tinkering, was also constitutive of a certain element of computer culture that relies on invention and innovation, two ideals that were lost in the transformation to a publicly marketable style. Bricolage becomes a system of "mass-marketed tinkering," by which anyone and everyone becomes a "hacker." It is the moment when hacker culture is commodified and, in the process, emerges as its opposite. Where bricolage originally was a way for the hacker to be close to the machine, tinkering with its various elements and operations, as a commodified form of software, bricolage serves to separate the user from the machine, effectively rendering the computer as an opaque object.

Within this system of representation, the computer, as an object, has been transformed from the ultimate transparent machine, built and programmed by its owner, into a "black box." As Jay David Bolter and Richard Grusin describe it, the process of technological innovation is always one of remediation, a process that follows a dual logic in which "our culture wants to both multiply its media and to erase all traces of mediation: ideally, it wants to erase its media in the very act of multiplying them."[5] As the computer undergoes the process of remediation, it becomes an increasingly opaque technology that is inaccessible and unable to be understood apart from the commodified "tinkering" that occurs on its surface. The commodification of bricolage has made it possible to experiment and tinker without understanding the object. As a commodity it is an inversion of the ideals that were essential to the subcultural representation itself. As Hebdige describes the process, once consumption becomes the central motor driving the cultural form, "the meanings attached to those commodities are purposefully distorted or overthrown."[6] Where, as exemplified by Greenblatt's chess program, hacking and bricolage represented the very possibility of complexity in the 1970s, by the 1980s they had come to represent simplicity. As a result, hackers have been gradually stripped of their "subversive power," as it has become increasingly difficult to "maintain any absolute distinction between commercial exploitation on the one hand and creativity/originality on the other, even though these categories are emphatically opposed in the value systems of most subcultures."[7]

The result of this incorporation can be seen both in the process of vilification of hackers and hacker culture by law enforcement and the judicial system[8] and in the spawning of new subcultures. In essence, even as it was perverted, hacker culture gained a measure of visibility. This initial system of computer style was linked to the machines that it produced, and it was that linkage to a system of material production that made its incorporation both possible and, arguably, inevitable. Hacker culture, as it would emerge in the underground of the 1980s and 1990s, would produce a new sense of style that would become divorced from the machinery that made it possible. In this process, hacker culture would undergo a stylistic transformation or mutation, making it both visible and highly resistant to incorporation.

Locating Hacker Style

With the production of visibility came the creation of a recognizable hacker style. This notion of style, however, is complicated by several factors. Unlike many identifiable forms of subcultural style, which often rely on physical attributes such as fashion, hacker style has manifested itself primarily in textual and electronic form, which is to say it is a subculture of information. From the 1980s to the present, hackers have developed a style that is suited to the digital medium. This electronic sense of style transcends the more traditional notions of style in several important respects. Electronic style is made possible by the transformation from a material to an information medium. That transformation means the primary site of production for subcultural style rests in the subculture itself, rather than in a "parent" culture. Because hacker subculture relies on information as the medium of representation, it is able to produce a style that is independent of the material elements of mainstream or dominant culture. The separation of production does not mean that there is no interaction between the subcultural and parent-culture systems of representations. Instead, it merely means that, as an information subculture, hackers maintain a higher degree of control over the means of the production of their own codes and systems of representation. From this notion of primary control flows the second implication of electronic style. With a premium placed on the fluidity of this style of representation, hacker subculture utilizes the more traditional notion of subcultural style as a means of resistance to incorporation. Hacker culture's ability to maintain control over a primary system of representation allows for the creation of a highly flexible and fluid process of resistance, which subverts efforts to incorporate, freeze, or integrate it.

Traditional subcultural style grows from what Hebdige has identified as "bricolage," the "science of the concrete," which is a system of "structured improvisation."[9] In essence, bricolage defines a style that reacts to contemporary culture by rearranging it, by reassembling elements of cultural significance (or insignificance) in ways that recode them, producing new meanings that are often subversive or represent reversals of commonly held beliefs. For hacker culture, bricolage is deployed as a *secondary* rather than *primary* strategy of resistance.

The anthropological origins of bricolage also illustrate precisely why we need to go beyond the traditional notion of both bricolage and subcultural style in understanding hacker culture. Bricolage, as originally conceived, presumes a certain disposition toward technology. Specifically, according to T. Hawkes, bricolage is situated by the "non-technical means" by which humans respond to the world around them.[10] In short, theories of subcultural style presume a relationship between technology and cultural codes that is thrown into question by hacker culture and hacker style. The grounding of style is shifted from the relationship of material to cultural production to one based on the production and consumption of knowledge. In the case of material and cultural production, dominant or mainstream culture both creates the material product and assigns it a cultural meaning and significance. Subcultural style is a *reaction* to and a reversal of those dominant cultural codes. As a redeployment of accepted and conventional meanings, bricolage functions as a re-arrangement designed to produce new meanings and "self-conscious commentaries" on issues of taste or style. Those rearrangements are prone to reabsorption into the cultural mainstream in a number of ways, but primarily through the process of "incorporation" — that is, through the reintegration of a new style as either a commodity form or an ideology.[11] Because of the relationship between cultural and material production, material subcultural style can always be appropriated and commodified. Such commodification serves to stabilize and "freeze" subcultural meanings in a way that deprives them of their subversive force (for example, punk fashion, such as safety pins, showing up on Paris runways). In contrast, because information is less constrained by material forces, it is able to remain fluid.

Representations of hackers and hacker style are, in contrast, a kind of technically savvy system of bricolage that does not need to wrest the originary moment of material production away from cultural interpretation. They are, instead, based on knowledge production rather than material production. As a result, hacker style is able to invent itself as an active system of signs rather than as a reactive system of rearticulation and recontextualization. In short, hacker style reverses the hierarchy of production. Specifically, hackers (especially old-school hackers) employ bricolage in the service of a new system of material production based on the reassembly of commonly or readily available material components, which are then *later* culturally coded by mainstream culture. Hackers transform knowledge into the moment of material production through concrete and structured improvisation, and it is mainstream culture that reinterprets that moment of material production into a cultural discourse. At that point (for example, when computers are themselves commodified), it is mainstream culture that continually reacts to hacker style. The effect appears to be the same, mainstream culture creating a dominant discourse and set of cultural codes that can then be made into commodities, but the difference in origin is significant. This difference in origin, coupled with the fact that hacker culture exists in a highly fluid electronic medium, means that hacker culture is extremely resistant to both commodity and ideological forms of incorporation.

In a second, perhaps more powerful, sense, hacker subculture relies on the fact that most people, even people who are considered computer-literate or computer experts by popular culture, actually have very little understanding of how their computers function. As a result, regardless of the degree to which hacker culture is commodified or incorporated, hackers still maintain a level of expertise over the machine and, as a result, over a particular dynamic. The roots of hacker subculture are in knowledge, particularly the knowledge that they understand how the systems that nonexperts are using function at a basic level. As a result, they are able to utilize, manipulate, and control those systems. The force of hacker subculture comes from the fact that, ironically, as it is commodified and incorporated as its opposite (a kind of surface tinkering that renders the technology increasingly opaque), it is continually increasing the gulf between hackers and end-users. Even as the system of commodification works to incorporate hackers' culture, it also opens up possibilities for hackers. It is precisely that gap in knowledge, expertise, and experience that hackers exploit in their endeavors.

Tendencies toward increasingly transparent interfaces are most commonly discussed in terms of the ways in which they make the technology more manageable, yet less accessible. What most analyses fail to examine is the space opened up between the expert and the end-user. In other words, most analyses look at the relationship to technology, not the relations between people that result from the technology. Hackers generally oppose such commodification and simplification but also recognize that the more layers that exist between the user and the machine, the more possibility exists to exploit that distance. Hackers' knowledge gains value as users become increasingly separated from their machines.

Put simply, the more hacker subculture becomes commodified and incorporated, the more opportunities it provides for hackers to exploit the fundamental misunderstandings that arise from that style of computer culture. It is a subculture that resists incorporation by turning incorporation into opportunity. For example, programs known as "Trojan horses" exploit such opportunities by attaching two programs to each other — one that provides a useful function, the other that does something surreptitious on the computer in a hidden or stealthy manner. A popular strategy for hacking BBSes in the 1980s relied extensively on these programs. A file would be uploaded to a system that would prove irresistible to the system operator. (This usually entailed pornographic images.) The system operator would run the program and watch a series of pornographic images, not realizing that, in the process, critical system files were being copied and mailed to various users' accounts. Program number 1 serves as a distraction for the real work of program 2, which is to gain information about the system, its users, and its secrets. The operation of Trojan horse programs relies precisely on the inability of the user to decipher what is happening on his or her own system. The easier it is to become a system administrator (or SysOp), the more likely it is that one will be vulnerable to such attacks.

The growth of the Internet has expanded the situation dramatically. Millions of people log on daily with little or no understanding of how their systems work or what risks or dangers can befall them as the result of a few mouse clicks. Even basic Windows functions such as file-sharing can give unauthorized access to a user's machine. Back doors and Trojan horses (such as Back Orifice) can give complete control of a Windows machine to a hacker without the user even knowing it.

Ease of access, particularly ease of access that masks complexity, makes end-users targets for hackers. Another popular ruse hackers use is to request that new users of IRC (Internet Relay Chat) type a series of commands that gives the hacker remote control over their user account. Because the command looks like gibberish to the end-user, and because the hacker appears to know what he or she is talking about, it was (and still is) extremely easy to compromise a UNIX account through IRC. In fact, most IRC servers post warnings telling users never to issue commands that they don't understand. Nevertheless, because a complexity exists that is masked by most popular IRC programs (such as mIRC), IRC is always open to exploitation because the very culture that made IRC possible through a difficult and complex system of tinkering has been commodified, stripped of that complexity, and placed on the surface. But in any such exchange, there is always a remainder, and that remainder, the space between expertise and end-users, is open to exploitation in a way that makes hacker culture continually able to renew itself and to exploit even its own commodification.

Styles of Resistance: Commodity and Ideological Incorporation

Hacker culture has proven incredibly resistant to most forms of incorporation. On the surface, it appears that hacker culture could be easily incorporated in terms of the computer. But it is important to remember that the material object of the machine has little (or nothing) to do with hacker culture itself. Thus, the incorporation of the computer as a machine has almost no impact on hacker subculture. Hacker subculture is about information.

The information about how the computer works is both increasing in complexity and growing more and more distant from the end-user. Ten years ago, it would have been impossible to use an IBM PC without at least a rudimentary understanding of how an operating system worked or the syntax of the operating-system command structure. Today, it is possible to do nearly everything you could do on the old IBM PC without touching the keyboard, much less understanding the command syntax or operating-system structure. While most users knew that DOS was a Disk Operating System, few users think of Windows 98 in those terms. Instead, Windows is seen as an interface, a bridge between the user and the computer to simplify its use.

In the rush to simplify computer interfaces, knowledge about complexity is impossible to commodify. Instead, the image of hackers is commodified in two different ways that actually reinforce their positions as computer experts. The first, most recent attempt at commodification comes in the form of positioning hackers as "threats" against which one needs protection. In 1999, IBM and Network Associates each ran TV commercials that depicted hackers breaking into corporate computer systems and causing havoc. In these ads, hackers were portrayed as outrageous figures who hacked to embarrass corporations by releasing corporate secrets, posting corporate salary information to the Internet, and generally breaking network security for the challenge of doing it.

Two messages become crystallized in these advertisements: first, hackers don't play by corporate rules — they exist outside the corporate mentality and, therefore, are unable or unlikely to be understood by the corporate mind-set; and, second, hackers present a real threat, one that corporate America has no idea of how to prepare for. The result is that companies like IBM and Network Associates are selling security as an "add-on" application for network servers and corporate computing.

This strategy further exoticizes computer hackers, making them seem increasingly mysterious and capable of almost superhuman hacking feats. Again, the split between end-users and experts becomes commodified, this time in the form of selling security. This form of commodification has taken a lesson from hackers themselves. It is a commodification that demands that hackers change and resist, and the more they change and resist, the more valuable these security services become. It is not surprising, in that respect, to find that most of these firms hire hackers to perform a great deal of security work and that hackers themselves have started their own security firms.

The second sense of commodification and incorporation is ideological. As Hebdige argues, this ideological maneuver results in either the trivialization, naturalization, or domestication of difference and is the means by which the other "can be transformed into meaningless exotica."[12] It is the process by which difference can either be naturalized or reduced to a meaningless diversion, not worthy of attention beyond amusement. There are a number of cases in which corporations (Microsoft in particular) have attempted to naturalize hacker exploits by claiming that software glitches that allowed hackers to break codes, steal passwords, and even take remote control of users' computers were, in fact, "features" and not bugs. Interestingly, a majority of Microsoft's press releases about security come from the marketing department, rather than programmers or software engineers. Attempts at naturalization have done little to alter what hackers do or how hackers are perceived. Instead, such attempts have generally angered hackers and made Microsoft more of a target than ever for the hacker community.

In fact, the only way in which hackers have been effectively "sold" is through a combination of the commodity and ideological forms of incorporation. It is hackers' exoticism that is sold. (For example, an IBM ad for security features a tattooed teen hacker with several prominent body piercings confessing to hacking for no discernible reason.) This exoticism is what renders hackers threatening, dangerous, and worthy of attention. Such a move makes their difference marketable. Difference is no longer a means of dismissal but is, instead, a warrant for attention and the reason we must pay attention. Hackers are a threat not only because we don't understand them, but also because they understand something important to us (our computers) in a way we don't. As a result, capitalist strategies of incorporation have found a way to rely upon, rather than break down, difference.

In that sense, we can argue that hacker subculture has maintained a powerful form of resistance. The incorporation that results from difference, although a very real form of incorporation, does not strip meaning from the subculture or in any way *determine* that subculture other than as an aftereffect. In essence, hacker subculture has found a way to subsist within a structure of incorporation that, instead of freezing, neutralizing, or dismantling it, relies on it to operate *as difference*. Industry, in that sense, relies on hacker culture's ability to invent and reinvent itself as a threat in order to sell security to consumers. The more sophisticated technology becomes, and the more reliant the user grows on it, the more important the figure of the hacker becomes. As a result, hacker culture has, at least to this point, survived alongside, even become enmeshed with, the corporate culture that incorporates it. It is an incorporation that sells difference without neutralizing it.

Where hackers do face problems of incorporation is in a medium as equally fluid as the electronic medium of the computer — that of popular culture. But because of the fluidity of each medium, the commodification of hacker culture has not functioned to "freeze" cultural or stylistic meanings but has, instead, spurred hacker culture to create different styles. As popular culture attempts to harness the electronic style of the hacker, it is the hackers who begin the process of commodification and incorporation within their own culture.

Hacking *Hackers:*
Incorporation and the Electronic Meaning of Style

Where *WarGames* spawned a generation of hackers in the mid-1980s, *Hackers* gave birth to a second generation of hackers in the 1990s. *Hackers* attracted the attention of a new generation of technologically literate hackers, who saw the Internet as the next frontier for exploration. The film spoke to this generation in the same way *WarGames* had spoken to earlier hackers, opening up a new world for exploration. Instead of focusing on the computer alone, *Hackers* introduced the new generation to the idea of a computer underground and the power of networked communication, things that older hackers had spent more than a decade building. Just as *WarGames* was the catalyst of the computer underground, *Hackers* sparked a second generation of hackers to follow the film's mantras: "Hack the Planet" and "Hackers of the World Unite." The messages of the two films, both deeply influential, were completely different and the subcultures that resulted from each would have very different perspectives.

Hackers tells the story of a group of high school students who find themselves caught up in a high-tech corporate swindle when the newest member of the group (a hacker so junior that he hasn't even been given a handle yet) copies a garbage file from the corporation's computer. The file contains evidence of the corporation's computer security expert's plan to steal millions of dollars from the company by releasing a computer worm that siphons funds in small, unnoticeable amounts from the system. The computer security expert is a former hacker who goes by the name The Plague, and when the group of hackers discover his plan, The Plague releases a computer virus in his own system that will cause worldwide ecological disaster. The Plague blames the hackers for the virus, and the gang is then pursued by Secret Service agent Richard Gill, a technophobic G-man, who is used by The Plague to track down the renegade hackers. Gill, who is little more than a puppet of the corporation, is shown to be a bumbling incompetent, unable to understand or navigate the world of technology he is supposed to be policing. The stage is set for a showdown of good versus evil, with the group of elite high school hackers taking on the evil computer genius, The Plague, to clear their names and save the world from ecological disaster. While the plot is pure Hollywood, the film's attention to detail about hackers and hacker culture helped it gain the attention of some hackers in the underground.

Of all the films about hackers, *Hackers* makes the most concerted effort to portray the hacker "scene" in some detail, even going so far as to get permission from Emmanuel Goldstein, the publisher of the hacker quarterly *2600,* to use his name for one of the characters in the film. In the film, hackers, for example, go by handles, rather than their real names; there is an abundance of references to the computer underground (some accurate, some wildly fantastic); and there is an effort to portray hacker culture as both an intellectual and a social system. These hackers are not isolated loners or misunderstood teens; they are cutting-edge techno-fetishists who live in a culture of "eliteness" defined by one's abilities to hack, phreak, and otherwise engage technological aspects of the world (including pirate TV and video games). The film, although making some effort to portray hackers realistically, is hyperbolic in its representations of hackers, law enforcement, technology, and underground culture. Accordingly, *Hackers* serves as a classic example of incorporation, the transformation of subcultural style into "commodity forms" (fashion, in particular) and into ideological forms (subcultural style refigured as "meaningless exotica").[13]

In contrast, hackers themselves have occasionally documented their own culture in an effort to resist media interpretations of their activities. One example of resistance is a film made by hackers themselves, showing the process of breaking into a telephone company control office and engaging in hacking practices. In it, two hackers, Minor Threat and Codec, enter the switching station of a telephone company and proceed to wreak havoc. During the course of their hacking exploits, they add free features to a friend's phone, examine the desks of several phone company employees, fake a call from a telephone company official, and explore various areas of the building. Mostly what the hackers do, however, is play. Nearly three-quarters of the tape is devoted to the hackers swinging on ceiling pipes like jungle gyms, playing hide and seek in the telephone switches, and skateboarding through the long, dark corridors of the building.

The film is a documentation of style, youth culture, and rebellion. There are continual comments about the phone company; pranks are played; and jokes and commentary are littered throughout the tape. The break-in and the subsequent activities are replete with the values of boy culture — expressions of hostility toward authority and constant statements of the hackers' own values and needs. It is a prototype for what Anthony Rotundo describes as an exciting way to attack the dignity of the adult world. In one of the last scenes of the film, the two hackers express their disapproval of the phone company (and arguably by extension the whole adult world) by relieving themselves on one of the phone company doors.

That film and *Hackers* are about essentially the same thing — hackers' relationship to corporate and capitalist dominance. The target in *Hackers* is a multinational corporation whose security expert plans to embezzle funds and plants a virus to cover his tracks. A group of hackers, the film's protagonists, are being set up to "take the fall" for the damage. The hackers' only hope is to outwit corporate and law enforcement intelligence and hack the "Gibson" computer, a machine considered impregnable by hackers and security experts alike. The challenge is only overcome by hackers banding together in a worldwide network — they enlist the aid of hackers all over the world by sending out an Internet distress signal.

Minor Threat's hacker film, like *Hackers,* also targets corporate America in perhaps its most idealized and bureaucratic form, the phone company. In this film, however, the hackers are neither wanted criminals nor imperiled innocents, wrongly accused. Instead, they are kids roaming around in a technological playground. The only moment of drama, a phone call from a security agent, whom they outwit with social-engineering tricks, is itself staged by the hackers. The hacking done in the film is — unlike that in *Hackers* — elementary, and there is no mystery and few risks (other than those manufactured by the hackers themselves) for the boys.

In comparing these two films, the differences in hacker style become apparent immediately. In Hollywood's version, hacker style is about hackers' relationship to technology, manifested in their clothing, their appearance, and, ultimately, their bodies. Accordingly, *Hackers* reduces hacker style to techno-fetishism. At one point in the film, the two protagonists (and ultimately love interests) have the following discussion:

[Dade typing at Kate's computer]
Kate: "It's too much machine for you."
Dade: "Yeah?"
Kate: "I hope you don't screw like you type."
Dade: "It has a killer refresh rate."
Kate: "P6 chip, triple the speed of the Pentium."
Dade: "Yeah. It's not just the chip. It has a PCI bus. But you knew that."
Kate: "Indeed. RISC architecture is going to change everything."
Dade: "Yeah, RISC is good. [Pauses, looking at Kate]. You sure this sweet machine is not going to waste?"

As a result of the conversation, Kate challenges Dade to a hacking contest. If she wins, he is forced to become "her slave"; if he wins, she has to go out with him on a date and, to add insult to injury, "wear a dress."

Kate's and Dade's discussion, technically, makes no sense. But as a matter of style it indicates the manner in which Hollywood film translates every aspect of hacker style, even the most basic social interactions, into technology. Throughout the film, all of the hackers'

Kate and Dade discuss computers and sex in the glow of the machine.

social interactions are mediated through and by technology. Early in the film, when the corporate evil hacker, The Plague (played by Fisher Stevens), needs to send a message to the film's protagonist, he does so by having a laptop delivered to him. When opened, the laptop projects an image of The Plague's face and issues a warning to Dade, threatening his family if he doesn't turn over a disk containing information about The Plague's plot to embezzle from the corporation for which he works. As fantastic as such an interaction may be, it illustrates precisely the way the film views hackers' interactions. Hackers are seen as figures who are only able to communicate about and through technology.

While communication in the film is conditioned by technology, there is an even more profound element that illustrates the manner in which technology dominates the narrative — the relationship of technology to the hacker's body.

Fashion as Style

In the film, hacker style is manifested in the wardrobe of the hackers. While several characters dress in typical teenage garb, the two lead hackers (played by Johnny Lee Miller and Angelina Jolie) prefer a high-tech vinyl and leather techno-fetish look. Miller's character (Dade Murphy, aka Zero Cool, aka Crash Override) and Jolie's character (Kate Libby, aka Acid Burn) serve as representatives of the hacker-elite sense of style. Their look is urban — very slick and ultracool. The female protagonist, Kate Libby, is every bit one of the guys, and even the eventual romance that develops between Kate and Dade is deferred until the final sequence of the film. Like most instantiations of boy culture, affection between Dade and Kate is displayed in a series of contests — first video games and later a series of hacks against Secret Service agent Richard Gill. In the process of creating mayhem (everything from ruining Gill's credit to having him pronounced dead), the contest becomes the means by which Dade and Kate express affection for each other, begrudgingly acknowledging each other's technical skills and cleverness.

While Kate's *character* is female, the *role* she plays is masculine, a hacker superior to everyone in her circle of friends, until she is challenged by the newcomer, Dade. Like the boys around her, all of Kate's sexual impulses are redirected toward technology. In fact, when a sexual encounter with her boyfriend is interrupted, she chooses to show off her new machine to her friends, rather than remain with her partner. The transformation is marked most clearly when the hackers, in the film's climax, engage in their final assault. Behind Kate is a sign signaling the message that has been made clear throughout the film, "Obey Your Technolust."

As the film progresses, the protagonists become increasingly enmeshed with technology. Initially, Dade interacts with his computer as a discrete machine; he types, and images and messages are reflected on the screen. The relationship is one with which we are familiar; the computer and user are separate and distinct entities. As the narrative moves forward and the hackers get closer to the prized data that they seek, their images begin to merge with the data itself. The image of a hacker's face is superimposed with the image of flying data, letters, numbers, and mathematical symbols. In the final scene, Dade no longer relies on either a computer screen or the merging of images but instead becomes physically integrated with the technology itself. For the final hacking scene, he wears an eyepiece, strapped

Hackers: From sunglasses to cyborg.

to his head, reminiscent of a kind of cyborg merger of man and machine. There is no longer a distance between Dade and the object of his hack; he has become the machine that he seeks to invade.

Fashion in *Hackers* is designed to mark the integration of humans and machines, allowing us to view hackers as technological creations. Thus, hacker style becomes integrated into a techno-fetishism that defines hackers and hacker culture by their relationship with technology and through the tools they use, the clothes they wear, and, eventually, the ways in which their bodies merge with data and machinery as they become completely absorbed in the machinery of hacking.

In contrast, hackers from Minor Threat's film behave in a manner that betrays a more common theme in hacker culture, that of the outlaw. Reminiscent of Jesse James and the Old West, these hackers wear bandannas to cover their faces and conceal their identities. Rather than hiding their identities through high-tech wizardry, they, instead, carefully put on latex gloves before touching the computer

A visit to the phone company: (*a*) playing the switch; (*b*) experimentation; and (*c*) adding a few features.

console they hack to avoid leaving fingerprints. They have taken on the traits of criminals, burglars, or high-tech espionage agents.

Unlike the characters in *Hackers,* these hackers engage with the technology in a decidedly more hands-on approach, offering a tour through the telephone switches and tinkering with different pieces of technology along the way. What is different about the hackers in this film is both the manner in which technology is represented and the distance that hackers keep from the objects they seek to invade. But Minor Threat's film can also be read as something more than a simple recording of a break-in to a telephone company control office. It is also a discourse to be read in opposition to the dominant media interpretation of hacker culture. The themes of the two films are so strikingly parallel that it is hard not to compare them at the levels of representation and reality. I want to resist such an impulse and suggest that the relationship between these two films is more complex than it might seem at first glance. *Hackers* did have a significant impact on the hacker community, whereas Minor Threat's film, which

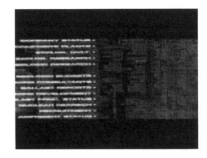

Images of technology: Hacking the "Gibson."

was routinely and secretly shown at hacker conventions, functioned much more to celebrate hacker culture than to document it.

Hackers was, by all accounts, a critical and commercial failure. Unlike WarGames, which captured national attention and the public imagination, Hackers was largely ignored. What makes Hackers important is the discourse that it put into circulation about the relationship between hackers and corporate culture. Where the discourse of WarGames was about the Cold War and the threat of nuclear annihilation, Hackers is about the 1990s discourse of global technology and capital and the rise (and power) of multinational corporations. It essentially explains why hackers resist corporate ideology, particularly that which can regulate or restrict access to information and communication (such as the ideology of the phone company). In short, Hackers is a discourse of justification for and, in part, explanation of what happens in Minor Threat's film. Hackers helps to explain why someone would break into a telephone company control office and exercise control over the switch.

Hacking the phone company: Terminals and COSMOS.

At that level, it is impossible to simply separate the representation of hackers offered in the film *Hackers* from the reality of hackers in Minor Threat's film. What *Hackers* illustrates is the threat of multi-national corporations, and what Minor Threat's film documents is the ability of hackers to intercede in that threat at the local level. In fact, to the new generation of hackers, Minor Threat's hacking demonstrates the promise and possibility of achieving a global hacking community (represented in *Hackers* by cuts to stereotypes of hackers all logging on: a Frenchman in a café, two Japanese hackers in ceremonial kimonos, and so on). The discourse of *Hackers,* at least in part, animates the film by Minor Threat. By having the larger context of corporate control and dominance read in *Hackers,* hackers are provided with a justification for targeting the phone company and are provided further justification for understanding hackers and hacking in relation to U.S. corporate culture.

Social Engineering

Even though *Hackers* was not a critical or commercial success, it did manage to attract the attention of hackers who were busy hacking, phreaking, and, generally, building their reputations. In response to what these hackers read as an unequivocal attempt to label, define, and commodify their culture, they accessed the film's Web site and redesigned it in protest. This was a strategy of resistance that has been gaining popularity in the computer underground.

Although most hackers dismiss hacking Web pages as prankish behavior, as not a "serious hack," it does remain an important part of the culture in a number of ways. First and foremost, as an act of resistance hacked Web pages serve as an expression of power in the most highly visible and increasingly commodified form of computer culture, the World Wide Web. These hacks range from the prankish to the political and from immature and racist to incredibly self-reflective.

MGM's Web page originally described the movie as a high-tech cyberthriller:

Zero Cool — real name Dade Murphy — is a legend among his peers. In 1988, he single-handedly crashed 1,507 computers on Wall Street, creating worldwide financial chaos.

Eleven years old, Dade Murphy had a record with the F.B.I. — forbidden to finger the keys of so much as a touch-tone phone until his 18th birthday, exiled from cyberspace. It's been seven years without a byte … and he's hungry.

Kate Libby, handle Acid Burn, has a souped-up laptop that can do 0 to 60 on the infobahn in a nanosecond. When Zero Cool collides with Acid Burn, the battle of the sexes goes into hard drive.

But all bets are off when they must pool their resources to battle The Plague, a master hacker employed by a corporate giant and using his considerable talents to worm his way into millions. Worse yet, he has hidden his own scheme by framing Dade, Kate and their friends in a diabolical industrial conspiracy. The young band of renegade hackers sets out to recruit the best of the cybernet underground to clear their names.

A fast-paced cyberpunk thriller, HACKERS delivers a wake-up call to today's computer generation concerning the enormous power at their fingertips. Director Iain Softley, who explored the roots of the '60s rock 'n' roll counterculture in Backbeat, now takes us behind the screens of the '90s computer subculture.[14]

The hackers responded by rearranging several images and posting messages that expressed their dissatisfaction with the way in which they were being represented. The hacked page bears the message (mimicking the lingo of film promotion):

THEY'VE LIED ABOUT US...ARRESTED US...
AND OUTLAWED US.
BUT THEY CAN'T KEEP US OUT...
AND THEY CAN'T SHUT US DOWN.

From that point on, the hackers trope on the industry lingo in an effort to rewrite the meaning of the film. Rather than a high-tech thriller, these hackers see the release of the film this way:

Hackers, the new action adventure movie from those idiots in Hollywood, takes you inside a world where there's no plot or creative thought, there's only boring rehashed ideas. Dade is a half-wit actor who's trying to fit into his new role. When a seriously righteous hacker uncovers MGM's plot to steal millions of dollars, Dade and his fellow "throwbacks of thespianism," Kate, Phreak, Cereal Killer and Lord Nikon, must face off against hordes of hackers, call in the FBI, and ponder a sinister UNIX patch called a "Trojan." Before it's over, Dade discovers his agent isn't taking his calls anymore, becomes the victim of a conspiracy, and falls in debt. All with the aid of his VISA card. Want the number?

What Kool-Aid was to Jonestown...What the 6502 is to the Cellular Telephone Industry...Hackers is to every Cyberpunk movie ever made. Check out the site and see for yourself.

There are two basic points of critique in this Web page hack. First, the hackers assert that the film is in some way unrepresentative of hacker culture and threatens to damage more serious or proper presentations of hacker culture. In fact, they argue that the film threatens to undermine an entire system of representations that describe hacker culture and with which hackers themselves identify. With the growth of the Internet, older hackers had already begun to witness an explosion of a new generation of hackers, who they regarded as uninformed and lazy. Unlike their generation, who had to explore and learn on their own, these new hackers, they feared, had it too easy. Terms such as "script kiddies," "lamers," and "wannabes" gained currency, describing hackers who had run sophisticated hacker programs ("scripts") without understanding them

Original Web page image for the film *Hackers* and modified image.

and had performed "lame" hacks without purpose or understanding. The hack occurred before the film's release, which accounts
for the hackers' dire predictions (Kool-Aid being the vehicle Jim
Jones's cultists used to commit mass suicide and 6502 CPU being the
processor that allowed hackers to hack and clone cellular phones).
Most of this older generation, themselves the product of *WarGames,*
understood the influence of media on hacker culture.

The argument is made with equal force through a hacking of one
of the film's title images, adding a carat and the word "Not" in front
of the title "Hackers." The hackers also further defaced the images
of the film's stars, scribbling over their faces and adding colorful
features. As a point of contrast, the hacked page displays a group of
hackers (taken at DEFCON, the yearly hacker convention) drinking
at a bar, their identities concealed by narrow black bands over their
eyes, calling up associations with victims who need to have their
identities protected.

A second critique embedded in the hack of the film's Web page
has to do with the very premise of the film. Those who hacked
the Web page argue that MGM (or any multinational corporation)
cannot make a film about hackers and global capitalism without
implicating itself. MGM's intent in making the film — "MGM's
plot to steal millions of dollars" — mirrors the film's essential argument. The hackers' point, then, is that MGM had to make a

"Real" hackers.

movie that followed its own ideology; it couldn't make a movie about real hackers because that would undermine its corporate viewpoint. MGM had too much at stake to describe in any accurate or reliable way what multinational capitalism is or hackers' relationship to multinationals.

The critique is as much a disavowal of the film's argument as it is a protest against unfairness or inaccuracy of representation. In that disavowal, however, there is also a reassertion of hackers' identity, of what it means to be a "real" hacker. At base, as is the case with most media and hackers, there is a contradiction. Hackers want the exposure, even if it only provides them the opportunity to critique it. Almost as if they could predict the impact on the next generation of hackers, the page closes with the following warning:

KNOWLEDGE ISN'T FREE
DON'T HACK THE PLANET
DON'T SEE HACKERS
IT SUCKS
BUY "TEACH YOURSELF C IN 21 DAYS" INSTEAD

The phrase "Hack the planet" is a mantra throughout *Hackers,* contrasting the movement of global capital and the globalization of technology, positioning the hacker at the global/local nexus. These hackers suggest two things: first, that things are much worse than they appear (knowledge isn't free), and, second, that the solution is not to be found in this (or any) movie, a lesson they had learned from their experience with *WarGames*. Hacking the planet isn't about going to see a movie; it's about learning how to program. The suggestion to buy "Teach Yourself C in 21 Days" is only half in jest. These hackers are concerned that the film's message of incorporation will be taken seriously by the current generation of "newbie" hackers, that they will see the globalization of technology and capital as the "liberation" of knowledge.

The concerns expressed by the hack, then, are not merely about accurate representation (although those concerns are clearly voiced). The hack touches upon a greater concern, a kind of vulnerability opened up by the possibility of incorporation by mass culture.

There is no doubt that hackers of the 1980s and hackers of the 1990s differ radically in a number of ways, not the least of which are their influences and inspirations. Hackers also recognize that the real danger to hacker culture comes from the enormous influence that mass culture has on the shaping of hacker identity.

Conclusions

Just as computer culture has undergone a radical shift through the process of incorporation, hacker culture has changed as well. While hackers have proven enormously resistant to incorporation from the computer industry, the influence of popular culture has been a more contested arena. Hacker identity is created and shaped by the split between a culture of expertise and a culture of end-users, but it is also heavily influenced and defined by images from popular culture, even among hackers themselves.

In examining how the industry and media shape the computer underground, several things become clear. First, the shift from a material to an information subculture has afforded hackers new strategies for reformulating subcultural identity and engaging in new strategies of resistance. Second, the threat of incorporation, traditionally structured as an economic and socially normative phenomenon, needs to be rethought around the thematic of information rather than material production. In keeping with that hypothesis, it is not surprising to find that the site of vulnerability for hacker culture and identity is not one of material production but one of information in the form of popular media and popular culture. Finally, it is not enough to understand hacker culture either as an entity separate from the representation of it or as a subculture formed exclusively in relation to those representations. Further, hacker culture, in shifting away from traditional norms of subculture formation, forces us to rethink the basic relationships between parent culture and subculture.

In what follows, I illustrate how this fusion of subcultural identity and its ambivalent relationship to material culture and parent culture problematizes the figure of the hacker in terms of broader cultural norms and representations, particularly in relation to the state, law, and law enforcement.

Part III

Hacking Law

Hacking Law

This part returns to the question of the representation of hackers in popular and, most important, judicial discourse to explore how hackers are defined both popularly and legally in terms of criminality. In tracing out the discourse of "computer crime," I argue that discourse about hackers' criminality is focused on issues of the body, addiction, and performance.

The domain of hackers is generally considered a "virtual space," a space without bodies. The technology of punishment, however, has its roots firmly established in mechanisms that relate almost exclusively to the body. Even surveillance, which takes the body and the visual as its target, is problematized by the notion of virtual space. As a result, hackers, as noncorporeal criminals, have corporeality juridically forced upon them. Crimes, such as trespassing, that occur in virtual space need to be documented in the physical world and attached to bodies. Attaching these crimes to the hackers' bodies does two things: first, it focuses on acts of possession rather than performance, and, second, it focuses attention on the hacker's body through metaphors of "hunting" and "violence."

Reading legal rulings regarding hackers and hacking and examining three cases of hackers who were "tracked" and, ultimately, captured and jailed, I illustrate the difficulty that the legal system has in the representation of hackers as criminals. Accordingly, it is the possession of secrets (such as passwords) that law prohibits, not the actual use of them. Criminality becomes defined by possession, rather than action. There are striking parallels between the discourse of computer crimes and that regarding illegal drugs. Hackers are often seen as "addicts," unable to control their compulsions — virtual beings who sacrifice their bodies to the drug of technology. The physical connection to technology is seen as the source of their criminality, as a performance of addiction, and courts have gone as far as banning hackers from touching computers, using Touch-Tone phones, or even being in the same room as a modem.

The second consequence of the juridical construction of hackers is the manner in which they are pursued. Describing hackers as "criminals," "terrorists," or even gang members connotes a sense of a threat directed toward the body through violence. The narratives of the pursuit and capture of Kevin Mitnick, Kevin Lee Poulsen, and members of the Legion of Doom and Masters of Deception are each read in relation to the questions of corporeality and the metaphors of the hunt and violence used to characterize them.

I conclude by examining the cases of two hackers, Kevin Mitnick and Chris Lamprecht, who found themselves confronted with rapidly changing standards — hacking had been transformed, in the eyes of law enforcement, from exploration and mischief into dangerous criminal behavior, resulting in new challenges to the law, questions of constitutional rights, a new era of political "hacktivism," and the political education of a new generation of hackers.

Chapter 6

Technology and Punishment: The Juridical Construction of the Hacker

> Hackers penetrate and ravage delicate private and publicly owned computer systems, infecting them with viruses and stealing sensitive materials for their own ends. These people...they're terrorists.
>
> — Richard Gill, *Hackers*

The image of Secret Service agent Richard Gill in the movie *Hackers* reflects both the hyperbole with which hackers are represented and the shallowness with which they are understood. Gill's description of hackers becomes something of a mantra, repeated continually throughout the film, making clear the fact that Gill, like many of the law enforcement officers he represents, has no idea what he is talking about. Instead, he is reciting a canned speech that is both sensationalistic and wildly inaccurate.

For all the hyperbole of Gill's statement, one aspect rings true — law enforcement's obsession with the corporeal. The highly sexualized metaphors of penetration and ravaging set against the delicacy of sensitive computers and data suggest that hackers are rapists and that computers are feminine. Further, this juxtapositioning makes a clear connection to the personification of information and makes it impossible to consider hackers solely in terms of the tools that they use. Technology, even to law enforcement, has become a problem of human relations, not merely a question of the tools that hackers use.

The separation of technology, conceived of as a problem of human relations, from the technical, formulated as a problem of instrumentality and utility, forces us to rethink several key aspects of the concepts of law and punishment. The problem has often been discussed as a technical one, presenting law and disorder on the electronic frontier as a "blend of high technology and outlaw culture."[1] There is a commonplace insistence, particularly in law enforcement, that there is nothing unusual about computer crime per se; what is unusual is the means by which hackers accomplish their tasks. To much of law enforcement, hackers are "common criminals" using uncommon means. Even defenders of hackers, such as Mitchell Kapor, argue for making a distinction. "Much of what is labeled 'computer crime,' " he proposes, "would constitute a crime regardless of the particular means of accomplishment. Theft of a lot of money funds through manipulation of computer accounts is grand theft. Does a computer make it any grander?"[2] The answer would appear to be "yes" considering the manner in which "computer crime" is described and detailed both in the media and in books documenting hackers' stories.

The intersection of technology and law problematizes and complicates what appears to be a simple and straightforward notion of "computer crime." Specifically, what is most commonly prosecuted in hacker cases is not the crime itself, what we might think of, for example, as electronic trespassing. Instead, the most common laws used to prosecute hacking cases are written around the possession of technology characterized as "unauthorized access devices." Taken literally, "computer crime" often means that the ownership of a computer or technology itself, or even the mere possession of it, constitutes a criminal offense. Such is the case with Ed Cummings (Bernie S.), who was incarcerated and held as a "danger to the community" for being in possession of a "red box" — a small, modified Radio Shack speed dialer that was altered to emit tones that would allow him to make free telephone calls from public pay phones. The first count of the indictment charged Cummings with "custody and control of a telecommunications instrument, that is a speed dialer, that had been modified and altered to obtain unauthorized use of telecommunication services through the use of public telephones." The second count alleged that Cummings had "custody and control of hardware and software, that is, an IBM 'Think Pad' laptop computer and computer disks, used for altering and modifying telecommunications instruments to obtain unauthorized access to telecommunications service."[3] In essence, Cummings was charged with possession of technology — possession of the computer itself became the crime.

It is important to note the fact that the possession of technology has become equated with and even been made to go beyond the performative act of its use. Here, the illegal performative act is *ownership* of the technology itself — not its use. Hacking, then, is constituted as a crime around the notion of the possession of technology. The reasons for this are simple. Hacking is an act that threatens a number of institutionally codified, regulated, and disciplined social mechanisms, not the least of which is the law. Hacking, as an activity, enters and exposes the foundational contradictions within the very structure of social existence. Two of those primary structures are the relationships between technology, law, and the body and the importance of the notion of secrecy in the operation of culture. In many ways, hacking at once performs and violates these central tenets. Hackers are continually using secrecy to reveal secrets and often find themselves in violation of the law without actually hacking. In this sense, hackers exist in a gray area where it is difficult to apply familiar standards of law and jurisdiction and hackers take delight in taunting law enforcement, trying to exist outside the law's reach.

Hacking and the Fear of Technology

At the most basic level, the reaction to hacking and hackers can be understood as a broader reaction to the threat of technology. However, it should be noted that within this dynamic, technology itself is not *precisely* what is feared. The tools themselves are actually completely benign. What is threatening, and what hackers and hacking expose, is the fact that as a relational concept that mediates the connection between and among people, technology is almost never benign. Part of the fascination with hacking has to do with what is not understood about it as an activity, and that lack of understanding is the source of genuine fear. As Katie Hafner and John Markoff explain it: "For many in this country, hackers have become the new magicians: they have mastered the machines that control modern life. This is a time of transition when young people are comfortable with a new technology that intimidates their elders. It's not surprising that parents, federal investigators, prosecutors and judges often panic when confronted with something they believe is too complicated to understand."[4]

This analysis is correct up to a point. However, to leave this misunderstanding of hacking as a technical problem, and to explain it as a response to a "fear of the new" or as a technophobic response to modern inventions, seems insufficient, particularly in terms of law enforcement and the judicial system. Lawyers and jurists confront things they don't understand on a daily basis. When doctors stand up to testify about spinal cord injuries, or lab technicians present DNA evidence, or engineers explain complicated issues in patent disputes, judges, lawyers, and prosecutors don't panic. The elements that need to be rethought in this dynamic are twofold. First, we must consider the relationship between cultural notions of technology and punishment. Second, we need to consider the problem of "computer crime" within the thematics that have been developed up to this point: anxiety over technology, displacement, and the culture of secrecy.

The problem of the relationship between technology and punishment has, in most analysis, primarily been conceived as a technical one. What is needed, however, is a more extended questioning of the problems of technology and punishment, which is to say, a reconceiving of the problem of punishment as a question of technology, as a question of human relations. This presents us with two lines of inquiry: What is the relationship between technology and punishment? In what ways and to what ends have various readings of technology, the technical, and punishment been deployed in representing, understanding, and misunderstanding hacker culture?

Understanding Technology and Punishment

In order to understand how hackers relate to the broader system of law, it is important to realize that they are positioned at, perhaps, the most central point for grasping how law and punishment intersect. Hackers, by understanding technology, are ideally situated to disrupt the most basic mechanism of social order — the relationship between technology and punishment. Following Nietzsche, Michel Foucault has done much to analyze the fundamental relationship between law, technology, and punishment. In *Discipline and Punish,* Foucault examines the history of modern incarceration, tracing out the various technologies that have made confinement and modern surveillance possible. Foucault sets out several guiding principles for understanding the relationships between technology and punishment. First, he writes, "Do not concentrate the study of punitive mechanisms on their 'repressive' effects alone, on their 'punishment' aspects alone, but situate them in a whole series of their possible positive effects, even if these seem marginal at first sight. As a consequence, regard punishment as a complex social function."[5] As a methodological precaution, this makes a great deal of sense, particularly in terms of the relationship between technology and punishment. If the goal of law enforcement is to "protect" us from high-tech hoodlums, as is so often claimed, the questions remain, What is it that is being "protected"? and What does it mean to be "protected"? and What are the "positive effects" of protection? Such questions seem easily answered at first sight, until one realizes that almost all of the high-profile cases that have been prosecuted do not involve common crimes. Hackers who enter systems and do nothing more than look around, or even copy files, do not profit from their crimes, generally do not do anything harmful or malicious, and do not cause any loss to the companies, organizations, or businesses that they intrude upon. Most often what hackers are accused of and prosecuted for is "trespassing" and "possession of unauthorized access devices." That is, they are prosecuted for their presence, virtual though it may be. Thus, the juridical system is protecting citizens not from the actions of hackers but from the presence, or the possibility of the presence, of hackers.

Here, we can identify one of the primary problems that confronts law enforcement: if presence is to be considered a crime, one needs something to be present. That presence can never be merely "virtual," but instead must be linked in some real manner to the physical world. In short, hacking needs a body. This body, however, cannot just be any body. It must be a body that has a particular call to the exercise of punishment, discipline, or regulation, which leads us to Foucault's second methodological premise: "Analyze punitive methods not simply as consequences of legislation or as indicators of social structures, but as techniques possessing their own specificity in the more general field of other ways of exercising power. Regard punishment as a political tactic."[6] In the case of the hacker, the technology of punishment must also be read not as just a technology of the body but as a politics of the body as well. We must understand that the virtual presence of the hacker is never enough to constitute crime — what is always needed is a body, a real body, a live body, through which law can institute its well-established exercises of power.

Third, Foucault insists that instead of "treating the history of penal law and the history of human sciences as two separate series," we should "make the technology of power the very principle both of the humanization of the penal system and of the knowledge of man."[7] As such, Foucault raises the issue of whether or not there is a "common matrix" or "a single process of 'epistemologico-juridical' formation" that gives rise to the questions of both law and human relations. If there is such a single element, then we ought to focus our attention on the commonality found in both the discussion of the law and human relations and particularly the points of intersection between the two. Nowhere in the discourse surrounding hacking is this more clear than in the manner in which hackers are characterized as criminals. Indeed, how exactly one characterizes hackers reveals a great deal about attitudes toward law and humanity. Hence, the discourse that surrounds the criminal nature of hacking reveals a great deal about broader social understandings of criminality, technology, and culture.

The Body and Memory

Technology has never been a stranger to punishment. It is easy to see how, particularly in the twentieth century, technology has, in fundamental ways, constituted punishment; it is one of the primary means by which power is deployed, networked, and regulated.

Technology, like all mechanisms of power, is also a means of resistance. As such, at moments, we are able to redefine and redeploy discourse and events in locally resistant ways. If we maintain that technology is a relational, rather than a technical, phenomenon, it becomes even more clear that it functions as one of the more complex networks of power. Relationships with technology infect every aspect of human communication, and technology mediates nearly every form of relationship. Even the most basic forms of face-to-face communication can be subject to recording, eavesdropping, or some other form of electronic snooping. In that sense, all human communication has at some level become public, insofar as all human relationships are mediated by and through technology that always threatens/promises to make that communication and those relationships public. Technology has become nothing more than the sum of and ordering of human relations that are in some manner mediated.

The connection between technology and punishment is as old as human civil relations. As Nietzsche argued, it is precisely the connection between technology and punishment that allowed human relations to become codified and regulated. One can read Nietzsche's *On the Genealogy of Morals,* perhaps one of the most insightful treatises on the relationship between culture and punishment, in just this light. Nietzsche insists on the separation of origins and utilities of things (particularly punishment), arguing that "the cause of the origin of a thing and its eventual utility, its actual employment and place in a system of purposes, lie worlds apart." Nowhere is this more clearly established than in his reading of the origins of punishment. Technologies of punishment evolved from the need to "create a memory in the human animal," and accordingly, the "answers and methods for solving this primeval problem were not precisely gentle."[8] What Nietzsche provides for us is the initial connection between human relations and technology: if one is to live in a civil society, one must follow certain rules, rules that are contrary to nature and even contrary to human survival. Hence, we find the very possibility of a civil society rooted in the technology of memory.

To call memory a technology is to suggest that it operates through a kind of mechanism that mediates human relations, and, as Nietzsche argues, punishment is precisely the technical mechanism by which we mediate all human relations through memory: "Man could never do without blood, torture, and sacrifices when he felt the need to create a memory for himself; the most dreadful sacrifices and pledges (sacrifices of the first-born among them), the most repulsive mutilations (castration, for example), the cruelest rites of all religious cults (and all religions are at the deepest level systems of cruelties) — all this has its origin in the instinct that realized that pain is the most powerful aid to mnemonics."[9] This sense of punishment, whatever else one can say about it, has been effective. The body, as we have seen time and time again, learns and remembers. The origin of punishment is creating a memory, not in the psychical consciousness but with and through the body as a physiological solution to the problem of memory.

There is a tendency to confuse the origin of punishment with the purpose of punishment. We tend to seek out a particular use or purpose for punishment (for example, deterrence, revenge, retribution) and ascribe that as the origin of punishment. In so doing, we mistake the technical utility for the technology itself. Punishment has a technical aspect, a means by which punishment is performed, which has as *its* goal a "purpose." But, as we have already seen, the basic technology that informs the understanding of the origin of punishment is the need to create a memory, *mnemotechnics*. While the latter, memory, serves to define and mediate human relations, the former, the infliction of pain, serves only instrumental and therefore technical ends.

Memory, as a technology, relies on the physical aspects of punishment. As a result, the technical aspect of punishment can never be divorced from the technology of memory. Punishment is, and always remains, *mnemotechnics* (literally, the combination of memory and technology). As Nietzsche states, "If something is to stay in the memory it must be burned in: only that which never ceases to *hurt* stays in the memory."[10]

Technology's relationship to punishment, however, does not end with this originary moment. Instead, technology has been the force that has propelled punishment, developing increasingly sophisticated ways to monitor, discipline, correct, and institutionalize norms, values, and ideals. This partnership between technology and punishment has always followed two basic principles: first, that it is the *body* and not the mind that remembers and, second, that the most powerful form of discipline comes not from an external implementation of coercion but when the subject of punishment actually incorporates the system of punishment into his or her own life and begins to do it to him or herself.

Most hackers understand both these principles. They understand that if the "crime" cannot be connected to a body, it cannot be punished. Moreover, they realize that, for the most part, the connection between technology and punishment is, at least in one sense, very tenuous. For the most part (and until recently), it was common for those who enforce the law to have little understanding of the technology that is used to break the law. As one hacker, Chris Goggans, describes his visit by federal agents: "So they continued on in the search of my house and when they found absolutely nothing having to do with computers, they started digging through other stuff. They found a bag of cable and wire and they decided they better take that, because I might be able to hook up my stereo, so they took that."[11] Indeed, law enforcement realizes that the information that will catch hackers and allow for their prosecution is not going to be computers, disks, or stereo cable, but instead the most valuable information will usually come from *other hackers.* Most hackers who are caught and/or sentenced are apprehended as a result of being turned in by another hacker. It is a hacker's relationship with other hackers, coupled with the threat of severe penalties for lack of cooperation, that provides information for most arrests.

One of the ways that law enforcement monitors hackers is by keeping careful watch on the relationships and networks that hackers set up among themselves. Evidence from and of these networks, usually the testimony of other hackers, is the most powerful evidence marshaled in criminal prosecutions of hackers. The most notorious hacker informer is Justin Tanner Peterson (Agent Steal), who worked for the FBI in the early 1990s, informing on dozens of hackers. However, in most cases, the principal evidence comes from hackers who have been bullied into cooperation by unusually high penalties and the threat of lengthy prison sentences.

There is an additional side to technology's relationship to the body that demands exploration. While it is the hacker's body that must be found, identified, and ultimately prosecuted, the relationship of hacking to the law has become curiously incorporeal in another sense.

The most common legal indictment against a hacker is "possession of counterfeit, unauthorized, and stolen access devices."[12] All this refers to passwords.[13] Access devices are items that allow one entry to a computer system and can be read as secrets that provide verification of identity. The parameters as to what constitute these access devices are written broadly (counterfeit, unauthorized, and stolen passwords would all count), but what is most remarkable about the law is that one never needs to use these access devices to be found guilty — all one needs to do is *possess* them. The constitutive act of possession is thus transformed juridically into the performative act of hacking. Legally, hacking ceases to be an activity and is reduced to involving only possession. All that needs to be proven is two things: first, that the hacker had the access devices in his or her possession and, second, that they are indeed "counterfeit, unauthorized, or stolen," which is to say that they have the quality of the secret.

Because these issues revolve around questions of identification, what is at stake is the juridical reconstitution of both the hacker and the subject. The premise of this law regarding access devices is that the information or the access device is, at some level, a secret shared between the system (which can authorize its use) and the user (who utilizes that system). That is, it expresses a relationship between them that is characterized by secrecy. But that relationship also produces a sense of identity. The performative act of sharing the secret, which occurs each time the user logs on, also betrays that user's identity. The system "knows" who is logging on precisely because that person (supposedly) is the only one who shares in the secret.[14] Without the quality of secrecy, each user could be anyone, and, therefore, the relationship would not perform any sort of identity. When a password is typed in, what the system does, in essence, is to verify the identity of the person on the other side — making sure they are not counterfeit or unauthorized. But that identity is, in all instances, a virtual identity, one that can be performed independently of the body. Identity, in this space, then, is reduced to a constitutive property — either I have the identity (for example, possess the secret) or I do not. Identity, in this virtual space, is severed from the body and, in that manner, becomes "performable" by another simply by knowing the secret. At this moment, the space between the performative and constitutive becomes undecidable — the act (if we can call it an act) of *knowing a secret* is indistinguishable from the act of *performing an identity.*

It is this separation of body and identity that makes the act of hacking possible and it is this separation that is taken up by the law. The hacker does not have to perform anything to violate this law. The only quality that the hacker needs to manifest is a constitutive one — proof that the hacker knows a secret that he or she is not supposed to know. How he or she got that information and whether or not it is used, which is to say performed, is of no consequence. The law itself affirms a crucial moment of secrecy, where simply knowing someone else's secret constitutes, legally, the performance of their identity.

Hacking the Panopticon: Reading Hacking as Resistance

Most hackers tend not to take law enforcement seriously, regarding their relationship as something of a game, a game that law enforcement is all too willing to play. Hackers commonly don FBI T-shirts and baseball caps and give themselves names that play on law enforcement (such as Agent Steal). Both sides realize, however, that just as hackers are trying to slip past law enforcement, the authorities are trying to catch hackers in the act. It is a game of watching and being watched. But, in all cases, what must be watched is the body, precisely the thing that is absent in the space of hacking.

Passwords constitute the most basic form of "unauthorized access devices," but there is a second function that these devices serve in relation to the body and identity. An "unauthorized access device" can also be considered a mechanism that masks the user's true identity. These devices, then, also create one secret as they mask or obscure another. One of the more common examples is the use of hacked cellular phones. These phones, which are altered to appear to belong to someone else (making their use both free and untraceable), serve to erect a barrier between the hacker and the network that they seek to enter. That barrier shields the hacker's body from the act of hacking. The system can monitor every act that the hacker performs, but it cannot locate the body that is performing the actions. The body, then, becomes the *secret* that the law must uncover, and the law against unauthorized access devices is aimed at precisely that secret. As a result, the law is positioned as protecting certain institutional secrets and preventing other "unauthorized" ones.

Hacking can be read in relation to this notion of a "culture of secrecy," which figures so prominently in the relationship between hackers and law enforcement. In what we can think of as an "incitement to discourse," the secret always plays a crucial role. It is the manner by which discourse can both be "confined to a shadow existence" and be "spoken of *ad infinitum.*" Foucault, for example, maintains that the *illusion* of the secret operates in such a way as to position its object *outside of discourse* and that only through a "breaking of a secret" can we "clear the way leading to it."[15] Put in different terms, access to the object of secrecy appears to be possible only through a breaking of the secret. This, however, is always only an illusion — it is the tension between the "shadow existence" and the "proliferation of discourse" that marks the force of the secret. As long as the object of secrecy appears outside of discourse, it can be talked about openly. Hackers, for instance, can speak openly about hacking, tools, and techniques and can boast proudly about systems they have entered. What they cannot, and for the most part will not, do is betray the information that gave them access. Hackers will rarely trade passwords or specific bugs, even with trusted colleagues, instead preferring to speak in generalities. The force of the secret remains intact as long as the hacker keeps particular pieces of information confidential.

This dynamic is steadfastly at work around issues of technology. The discourse around technology has exploded in the past decade, particularly with the growth of the Internet. What is talked about, in terms of hackers at least, is the manner in which hackers exist in a shadowy space of secrecy, possessing near-mystical powers that allow control of technology that itself is beyond discourse. The hackers are coded in such a way that they become the secret that needs to be broken. The discourse surrounding hacking reveals little about hackers themselves; instead, it tells us a great deal about social attitudes toward technology.

This secrecy is regulated by a system of panopticism, an all-seeing system of surveillance that serves to instill a sense of always being watched in those who are subjected to it. Panopticism generates its power not by continually monitoring but by making it impossible to determine if you are being monitored at any given time. As a result, it is always possible that you are being watched, and it is impossible to determine with certainty if you are not being watched. It is the process by which subjects learn to govern and discipline themselves, internalizing the thought that they are continually under observation. The structure of panopticism exploits the secret in two ways through the dynamic of surveillance. Those who monitor do so by exploiting a secret — whether or not one is being watched. The secret, then, whether or not someone is actually monitoring, preserves the power of the panoptical gaze. That gaze, however, is aimed precisely at the notion of secrecy itself. Panopticism's goal is the complete removal of the space where secrecy can operate — ideally, in a panoptical space, no one operates in secret because it is always possible that one's actions are being watched. What is watched, and this is of crucial importance, is the body and the space that the body occupies. In terms of hackers, however, that body and that space are rendered "unwatchable." One can watch a hacker's actions, even monitor them, online, but this means nothing until they can be attached to a real body and therefore are prosecutable.

The hacker can be read in this respect as a figure who both deploys and disturbs the notion of the secret, particularly in relation to the law. In short, hackers' use of the secret is made possible by the space that the broader culture of secrecy opens up. The law targets the connection of the secret to the hacker's body. Simply finding the body is not enough — the law must attach the body to a secret. In one case in March of 1990, Chris Goggans was raided by Secret Service agents in just such a quest. As he recalls the events, Agent Foley approached him after a thorough search of the premises and confronted him with some business cards he had made up that read "Eric Bloodaxe, Hacker," along with a small U.S. Treasury logo. Goggans responded to Foley: "Well, it doesn't say anywhere on there 'Chris Goggans, Special Agent.' It says, 'Eric Bloodaxe, Hacker.' Whoever this Erik Bloodaxe character is. It might be me, it might not. I'm Chris Goggans and that says 'Eric Bloodaxe, Hacker.' " The connection would not be enough to convict Goggans of anything; instead, Foley insists that there must be some secret to be told and he utilizes the threat of the law in an effort to extract that secret. As Goggans explains: "He says, 'Well if you don't tell us everything that there is to know about your higher ups, we are going to be pressing state, local, and federal charges against you.' I said, 'On what grounds?' He goes, 'We want to know everything about your higher ups.' Which I'm thinking, gosh, I'm going to have to turn in the big man, which is ludicrous, because there is no such thing as a higher up, but apparently they thought we were a part of some big organization."[16] Failing to connect Goggans to the secret physically, law enforcement tried to connect other bodies to Goggans through discourse. The demand was for Goggans to break the secret — "Tell us everything that there is to know" — and to connect other bodies to other secrets.

Addiction and Technology:
Rethinking the Cyberbody

The body is the locus of criminality and deviance, as well as punishment, justice, and correction. It is identifiable, definable, and confinable. Taking up the mantle of cyberpunk science fiction, hacking envisions a world without bodies, in which hackers exist, first and foremost, as virtual beings. Such an incorporeal nature is generally thought of as a technical invention, perhaps best described by William Gibson in his envisioning of cyberspace in his 1984 novel, *Neuromancer.*[17] In *Neuromancer,* Gibson tells the story of Case, a computer cowboy, who, after stealing from his employer, was neurologically damaged as a form of punishment or payback, damage that made his body no longer capable of interfacing with the computer matrix. Describing the protagonist's now-defunct relationship to the technological, Gibson writes: "For Case, who'd lived for the bodiless exultation of cyberspace, it was the Fall. In the bars he'd frequented as a cowboy hotshot, the elite stance involved a certain relaxed contempt for the flesh. The body was meat. Case fell into a prison of his own flesh."[18] The primary vision of hacking, then, is founded in the hacker's reliance upon the technological. The infliction of such punishment is not confined, however, to the world of the future. In the everyday world of hacking and "computer crime," the elimination of the technological is the greatest threat the hacker faces, and, not unlike Case's employers, judges are fond of proscribing penalties for hackers that include forbidding them to access technology such as telephones, computers, or modems.[19] The modern judicial system attempts to legally produce the equivalent of Case's neurological damage.

The 1988 arrest, trial, and conviction of Kevin Mitnick for breaking into the phone company's COSMOS system (the computer system that controls phone service) provide a striking parallel to Gibson's character Case. During the trial itself, the judge "sharply restricted his telephone access," allowing Mitnick to call only "those numbers that had been approved by the court." After Mitnick was found guilty (and served prison time), his relationship to the technological was diagnosed as "compulsive," and after his release he was prohibited from touching computers. A short time after, when it was determined that he could control his behavior, Mitnick was allowed to use computers again and even to look for employment in computer-related fields, but he was still not allowed to use a modem.[20]

Even more striking are the conditions of probation for Kevin Poulsen, another Los Angeles hacker. Poulsen was convicted of fraud for using his computer to illegally fix radio call-in contests (among other things) and was given the following "special conditions" of supervision for probation:

[Y]ou shall not obtain or possess any driver's license, social security number, birth certificate, passport or any other form of identification without the prior approval of the probation officer and further, you shall not use for any purpose or in any manner, any name other than your legal true name; you shall not obtain or possess any computer or computer related equipment or programs without the permission and approval of the probation officer; and you shall not seek or maintain employment that allows you access to computer equipment without prior approval of the probation officer.[21]

Poulsen wrote, "It got even more interesting when I was released. When I reported to my P.O., he explained to me that, not only could I not use any computer, with or without a modem, but that I couldn't be in the same room as a computer. I had to look for a job with an employer that had no computer equipment on the premises. 'Oh, and by the way, don't forget that you have to pay $65,000.00 in restitution in the next three years.' "[22]

Characterizations of compulsive behavior were employed not only by the courts. The notion of addiction was used by Mitnick's lawyers in an effort to get a reduced sentence. After Mitnick's arrest in 1988, his "lawyer convinced the judge that Mr. Mitnick's problem was similar to a drug or gambling addiction."[23] After his release, Mitnick was sentenced to six months in a halfway house, complete with a twelve-step program for drug and alcohol offenders.

The notion of addiction, particularly in Mitnick's case, is specifically located in terms of the body. As Hafner and Markoff describe him, Mitnick was "plump and bespeckled," "the kind of kid who would be picked last for the school team," and "his pear-shaped body was so irregular that any pair of blue jeans would be an imperfect fit."[24] In almost all accounts, his body is written as the *cause* of his addiction. Harriet Rossetto, his counselor from the center in Los Angeles where Mitnick underwent his treatment, attributes his addiction to computers to the fact that "he is an overweight computer nerd, but when he is behind a keyboard he feels omnipotent."[25] Even John Markoff, a staff writer for the *New York Times* who had followed Mitnick's story for a number of years, described an almost involuntary relationship between technology and Mitnick's body. "During the treatment program," Markoff writes, "Mr. Mitnick was prohibited from touching a computer or modem. He began exercising regularly and lost more than 100 pounds."[26] Markoff and others seem to suggest that it is the physical connection to technology that perverts and deforms the body. Joshua Quittner, writing for *Time*, reports the connection in precisely the same way: "As a condition of his release from jail in 1990, he was ordered not to touch a computer or modem. By June of 1992 he was working for a private eye, doing surveillance and research, and had dropped 100 lbs."[27] The connection between Mitnick's not "touching" computers and modems and his weight loss is presented as mininarrative in and of itself, a narrative that suggests both a causal connection between his lack of access to technology and his weight as well as the broader notion that technology is somehow harmful to the body. While the first connection is obvious on the face of things, the second is a bit more elusive.

Hacking, according to the judicial system, is akin to "substance abuse" (the actual term deployed by Mariana Pfaelzer, the sentencing judge for the U.S. district court in Mitnick's case). The judge's decision was the result of the tactics of Mitnick's attorney, who argued "that his client's computer behavior was something over which his client had little control, not unlike the compulsion to take drugs, drink alcohol or shoplift." As a result, Mitnick was sentenced to a one-year prison term with six months of rehabilitation to be served in a halfway house. Mitnick continued attending meetings for codependent children and children of alcoholics following his release from the halfway house.[28]

What is interesting about this treatment is the manner in which the law and the structures of punishment remain blind to the social dimensions of technology. In Mitnick's case, the computer is viewed as an object that is essentially negative in character. It is not a value-neutral tool, one that can be used beneficially or maliciously. It is positioned not as a substance but as a dangerous substance. Computers are likened to drugs and alcohol. The shift is subtle but important, and it betrays an underlying anxiety and hostility toward technology. It is also, most likely, the reason why the plea was successful.

The problematic nature of drugs has centered (at least since Plato's time) on the undecidability of their nature. Drugs have traditionally been regarded as "substances" that, when taken, have the ability either to poison or to cure.[29] Particularly in the wake of national hysteria, including the "War on Drugs" and the Reagan administration's "Just Say No" campaigns, drugs have taken on a fundamentally negative symbolic valance. There are at least two reasons for this transformation. First, the national campaigns dating from the mid-1970s and 1980s have heavily coded drugs as dangerous, deadly, and addictive. Second, in the wake of HIV and AIDS, drugs have begun to lose their positive valance. HIV has created a rhetoric of viral infection that has rendered the positive symbol of drugs impotent. These two movements — the first signifying the destructive, essentially negative, power of drugs, the second, the medicinal impotence of healing drugs — have had the effect of rendering the undecidable nature of drugs decidable.

The equation of technology with drug addiction is a powerful one. It is also the means by which technology is attached to the body and out of which is constructed the activity of hacking not as a malicious or even intentional activity but, instead, as an obsessive disorder resulting from physical contact with the object of the obsession. Mitnick's "treatment" consists of not "touching" a computer or modem, suggesting that it is the physical contact with technology, rather than the actual usage of it, that produces the addiction. Again, the body, particularly its physicality, plays a crucial role in the construction of the hacker. Technology itself is written as a drug, and the hacker is written as an addict. As Mike Godwin of the Electronic Frontier Foundation puts it, "The great ones are all obsessed, which is what it's about."[30]

As a result of the discourse surrounding hacking and the body, the most interesting definitions of computer crime are to be found in the nomenclature used to describe hackers during actual investigations and "manhunts." The tracking of hackers is a discourse so thoroughly gendered that it is impossible to separate the characterizations of the hunter and his prey from traditional masculine stereotypes. The two enact a drama of the hunter and the hunted — a contest of wills, where one will emerge victorious and one will be defeated. The hunter tracks the hacker, we are led to believe, using only his wits, cunning, and instinct. It is an act that reduces each to their most primitive, masculine roles. Just as the hunter relies on his instincts to bring in his quarry, the hunted survives on his abilities to escape detection and foil the hunter's efforts. The discourse surrounding the hunt leads us to believe that the contest is decided, fundamentally, by who has the better instincts or who is the better man.

It is during these "hunts" that the characterizations of the "cyberbody" take on heightened importance and emphasis. During these periods of pursuit, there is a mechanism for describing hackers that deploys a well-embedded narrative that fosters clear perceptions of who the hacker is and what threat she or he poses. It is commonly framed in the basic "cops and robbers" vernacular, where the hacker is often described in criminal, but nonviolent, terms. Those who pursue him are often characterized as "sleuths," "trackers," or "hunters." The hacker is a "cyberthief" or, for high drama, a "master cyberthief," and the pursuit invokes the language of the hunt ("tracking," "snaring," "tracing," or "retracing"). Often this narrative will feature hackers as "fugitives" who "elude," and, repeatedly, they will be described (once apprehended) as being "caught in their own web." In some cases, hackers are even given honorific titles such as "Prince of Hackers" or "Break-in Artist."

The most compelling aspect of these narratives is the manner in which the metaphors of the hunt are enacted. The hunt is not, as one might immediately suspect, a strategy of depersonalization — the hacker is not reduced to some animal form that is tracked, hunted, and captured or killed. In fact, the discourse of the hacker is less about the hunted and more about the hunter. As we read about the hunt, we uncover two dynamics: first, the drama of the hunt, which always seems to hold a particular narrative fascination; and, second, the narrative of the hunter, who, in order to catch his or her prey, must learn to think like them. Part of our fascination is with the act of repetition, which we live out vicariously, through the hunt. We watch as the hunter learns to think like the hunted, and it is through *that* process that information about the hacker's motives, intentions, and worldview is disclosed.

The drama of the hunt is an extremely familiar narrative, and it presumes a great deal of information: the activity is adversarial; there is are hunter, a hunted, and a gamelike quality that relies on deception, sleight of hand, illusion, and misdirection. Indeed, the notion of a "trap" relies, at its heart, on the act of deception. It must look like something appealing, when in fact it presents great danger. The manner in which the hunt is described, and often enacted, is also about how one thinks, particularly how the hunted thinks. Will he or she be smart enough to see the trap in advance? Will he or she outsmart the hunter? The structure of the hunt, then, is primarily a battle of intelligence, a test of skill, rather than one of physicality or even will.

Accordingly, the hunt is always about thinking the thoughts of the other. If the hunter is to succeed, he or she must understand the hunted better than the hunted understands the hunter. As a result, the hunt begins to exist in a world of its own, a world that possesses a gamelike quality. John Markoff, after helping to get Mitnick arrested, explains his reaction to seeing him being sent to jail: "It felt odd to me. It was as if it had all been a game, and all of a sudden the game was over and everybody realized this is the real world."[31] The sense of "real worldness" can be traced to the moment when federal agents knocked on Kevin Mitnick's door and placed him under arrest. The game lost its gamelike quality the moment a body was made present. Even Markoff, who had covered Mitnick's case before, as well as a host of other computer stories in Silicon Valley, fell into the spirit of the game, playing along even as he reported the story for the *New York Times*.

In this sense, we must also understand the hunt as a pursuit of the body. In the case of Kevin Mitnick, everything that was needed to make the arrest and prosecute the case was already known, documented, and recorded. Given the manner in which Mitnick's body figures into the narrative of his relationship to technology, it shouldn't be surprising that his "Wanted" poster reads, under the heading "Miscellaneous Information" — "Subject suffers from a weight problem and may have experienced weight gain or weight loss."[32]

The case of Kevin Mitnick is perhaps the clearest account of the ways in which bodies are constructed in the discourse of hacking and law enforcement. There are, however, a number of other stories, each of which tells the story of a hunt. In each case, these hunts are also moments of displacement of fears or anxieties that have become attached to the threat of technology and its intersection with everyday life.

Three Hunts in Cyberspace

Of the many cases that illustrate the relationships among hackers, technology, and punishment, three in particular demand special attention because of their media exposure and high profile. Each of these cases is unusual, in part, because the subjects of them received jail time, an unusual penalty for hacking offenses, particularly for young adults. The first, the story of Kevin Mitnick, has been the subject of at least three recent books and tells the tale of federal agents, who were led on a series of cross-country manhunts, and of Mitnick, who went by the handle "Condor," finally being apprehended by Tsutomu Shimomura, a computer security expert at the NSCA center in San Diego. The second, the story of hacker Mark Abene, also known as Phiber Optik, explores the underground war between rival hacker groups, the LOD (Legion of Doom) and the MOD (Masters of Deception). In the aftermath, Abene would be charged with a series of crimes and end up spending more than two years in jail. The third story is about Kevin Poulsen, a Los Angeles hacker who was arrested after being featured on *Unsolved Mysteries,* a TV program that helps law enforcement capture fugitives from justice. What is unusual about these three cases is not what the hackers themselves did but rather the reactions that their actions prompted and the manner in which each was described, reported, and detailed in the popular press and media.

Kevin Mitnick: The Hunt for the Body

February 15, 1995. At 2:00 A.M. federal agents knocked on the door of apartment 202 of the Players Court apartment complex in Raleigh, North Carolina. When Kevin D. Mitnick answered the door, he was taken into custody, ending a three-year search for "one of the most wanted computer criminals" and what U.S. state's attorney Kent Walker called "a very big threat."[33]

Mitnick's reputation as an *überhacker* is legendary and gives insight as to why both media and law enforcement gave particular attention to his case. Beginning with his ability to manipulate phone switching systems in the early 1980s, Mitnick's talents allowed him to engage in a wide range of exploits, from annoying nuisances (such as using computers to change people's telephones to pay phones, making them unable to dial without first depositing money) to pranks (such as intercepting 411 [directory assistance] calls or rerouting calls to unsuspecting phone numbers) to grand larceny (stealing the entire DEC VMS operating system while security experts at Digital simply watched millions of lines of their code being downloaded off their machine, unable to do anything about it).[34] It was the latter that would get Mitnick arrested, prosecuted, and convicted for the first time in 1989. Mitnick's social-engineering and phreaking abilities were always the bedrock of his hacks. Calls he made were always untraceable because he had either manipulated the phone company software to hide his tracks or had simply found a way to bounce the call from one phone system to another, making it appear to each that the call had originated from the other. In the early 1990s, Mitnick would run into trouble with the law again. The hack that made him a fugitive for a second time was a wiretap that he had installed to allow him to listen in and gain the access codes of FBI agents as they called into the California Department of Motor Vehicles.[35] These codes allowed Mitnick access to the entire driver's license database for California. According to authorities, this violation, his failure to report to his probation officer, and Mitnick's reported parole violation (he was not allowed to touch computers or a modem while on parole) led him to drop out of sight and prompted the three-year search that would ultimately lead to his arrest. Mitnick, for his part, denies that he was ever given notice of a violation or failure to report and simply left town at the end of his probation period (an argument that was confirmed when his status as "fugitive" was retroactively overturned by the courts).

The Hunt

Mitnick's arrest was the result, in part, of the work and technical expertise of computer security expert Tsutomu Shimomura. Shimomura believes that on Christmas day, 1994, Mitnick hacked his system and downloaded Internet security programs as well as programs designed to hack cellular telephone equipment. Shimomura's investigation (as well as several prank phone calls allegedly from Mitnick himself) revealed Mitnick's identity. From that point on, Shimomura began his own personal manhunt (with the help of the FBI, local telephone companies, long-distance service providers, colleagues from the San Diego Supercomputer Center, and independent computer consultants), which led them first to The Well, a Sausalitobased Internet service provider where Mitnick had stashed the files he had downloaded from Shimomura's machine. From there, the trail led to Netcom, the San Jose provider that had recently had its more than twenty thousand credit card numbers lifted by someone who was assumed to be Mitnick (even though the file of credit card numbers, which Netcom had stored online, unencrypted, had been circulating on the Internet for months prior to Mitnick's use of the system). At this point federal investigators had pinned down Mitnick's location. They were certain he was operating from somewhere in Colorado (almost two thousand miles away from his actual center of operations). Shimomura was able to identify two other points of operation — Minneapolis and Raleigh, each of which had Netcom dial-in numbers. As telephone records were searched, investigators were able to narrow the search to Raleigh, where calls were being made with a cellular modem. Calls "were moving through a local switching office operated by GTE Corp. But GTE's records showed that the calls looped through a nearby cellular phone switch operated by Sprint.... Neither company had a record identifying the cellular phone."[36]

By using cellular tracking equipment, investigators (primarily Sprint technicians) were able to locate the building and eventually the apartment from which Mitnick was operating. As Markoff told the story in the *New York Times:* "On Tuesday evening, the agents had an address — Apartment 202 — and at 8:30 P.M. a federal judge in Raleigh issued the warrant from his home. At 2:00 A.M. Wednesday, while a cold rain fell in Raleigh, FBI agents knocked on the door of Apartment 202. It took Mitnick more than five minutes to open it. When he did, he said he was on the phone with his lawyer. But when an agent took the receiver, the line went dead." Subsequently, Mitnick has been charged with two federal crimes: "illegal use of a telephone access device, punishable by up to 15 years in prison and a $250,000 fine," and "computer fraud, [which] carries penalties of 20 years in prison and a $250,000 fine."[37] The story of Mitnick's capture and arrest has been chronicled in a host of newspaper articles and magazine stories and recently in two books, *Takedown: The Pursuit and Capture of Kevin Mitnick, America's Most Wanted Computer Outlaw — by the Man Who Did It* (by Tsutomu Shimomura, with John Markoff) and *The Fugitive Game: Online with Kevin Mitnick* (by Jonathan Littman).

The manner in which Mitnick's arrest was chronicled reveals a great deal about popular perceptions of hackers. Of the various hackers who have been "hunted" by law enforcement, Mitnick is by far the clearest example of the metaphor's enactment. Mitnick, however, was not the usual criminal; he wasn't even the usual cybercriminal. Shimomura, the security expert who would eventually find Mitnick, was hired by two Internet companies, The Well and Netcom, to catch Mitnick. As Jonathan Littman reports, because Shimomura was considered a "civilian" by law enforcement personnel, they wanted to exclude him from the final search of Mitnick's apartment.[38] The enactment of the "hunt" metaphor was a hunt for a body, Mitnick's body, and manner in which that hunt would be accomplished would be a battle of skills between Mitnick and Shimomura. But it was also the story of a boy war, reported by Markoff as a test of masculinity and skill.

In Shimomura and Markoff's telling of the incident, the contest between the two men was a "battle of values," where Shimomura represented the "honorable samurai" and Mitnick the "evil genius." But the hunt itself was also a game or contest. The two would not meet face-to-face until after Mitnick's arrest. Shimomura describes the encounter this way: "Halfway into the room he recognized us and paused for a moment. He appeared stunned, and his eyes went wide. 'You're Tsutomu!' he said, with surprise in his voice, and then he looked at the reporter sitting next to me. 'And you're Markoff.' Both of us nodded."[39]

The intervention of the law into this contest or game radically transformed it for both the hunter and the hunted. Once the contest was rendered corporeal, the stakes were immediately changed. As Shimomura realized, "It had become clear to both Mitnick and me that this was no longer a game. I had thought of the chase and the capture as sport, but it was now apparent that it was quite real and had real consequences."[40]

Although Mitnick had never profited from any of his hacks and he had never deprived anyone of data, information, or service, he was charged with telecommunications fraud and computer fraud, each of which carried a possible sentence of fifteen years, charges only made possible by the connection of Mitnick to computer intrusion by physical evidence. Where Shimomura had seen only a contest in virtual space between himself and a hacker, law enforcement had seen criminality. Even as Shimomura considered Mitnick to be "petty and vindictive," guilty of "invasion of other people's privacy" and "pursuit of their intellectual property," he still remained ambivalent about Mitnick's arrest— "Strangely," he writes, "I felt neither good or bad about seeing him on his way to jail, just vaguely unsatisfied. It wasn't an elegant solution — not because I bought some people's claims that Mitnick was someone innocently exploring cyberspace, without even the white-collar criminal's profit motive, but because he seemed to be a special case in so many ways."[41]

In many ways, Shimomura's analysis details precisely the crime of which Mitnick was guilty — a crime of identity, the crime of being Kevin Mitnick. The threat that Mitnick posed was also the thing that made him so difficult to track down and capture. It was a case of fraud in which no one was defrauded, in which nothing of value was taken or destroyed, and which likely would have been entirely a noncriminal matter had Mitnick, in fact, been someone else.

What the case of Mitnick makes clear is that the criminal dimension of hacking is entirely dependent on the connection of a virtual identity to a corporeal presence who is anyone other than who they claim to be. This virtual/corporeal split is what animates the metaphor of the hunt and what demands that the hacker's body be the subject of law and representation in juridical discourse. The body that both law enforcement and the media chased, however, was a body that had been in large part the invention of the Southern California media.

Creating Kevin: The Darkside Hacker and the Southern California Media (From Los Angeles to New York and Back Again)

Remember, I didn't make up the term "dark-side hacker"; that was an invention of the Southern California press.

— John Markoff

Mitnick's story has been told, it seems, by just about everyone but Mitnick himself. The story, which began as a local-interest piece on the pages of the San Fernando Valley's *Daily News,* eventually went national (being featured as a page-one story in the *New York Times*) and returned to Hollywood as the subject of a Miramax film. The first portrayals of Mitnick were inventions of the Los Angeles media, which portrayed him as a dangerous criminal able to perform miraculous feats with phones and computers. After his initial apprehension, the *Daily News* wrote several stories essentially retracting many of the claims it had made about Mitnick earlier, but those stories were not picked up by the national press. The one label that did stick was "darkside hacker." Thus Mitnick, who had never profited from his hacking, nor done any damage to computer systems, files, or code, was branded as "darkside hacker" by the local media. In 1989, the *Los Angeles Times* set the tone for how Mitnick's story would be covered. The piece, which ran January 8, displayed the headline " 'Dark Side' Hacker Seen as Electronic Terrorist," a headline that played upon a range of cultural anxieties and cultural icons made popular by the film trilogy *Star Wars.* In response to local media attention, *USA Today* ran a front-page story about Mitnick that included an image of Kevin Mitnick's head, morphed with Darth Vader's mask and body, graphically illustrating Mitnick's conversion to the forces of evil.

The second, and perhaps the most important, portrayal of Mitnick occurred in a book by Katie Hafner and John Markoff, titled *Cyberpunk,* in which Mitnick was portrayed as a "darkside hacker" at length. The same label that had led *USA Today* to publish a picture of Mitnick's face superimposed over an image of Darth Vader proved a powerful hook for drawing the reader's attention into the Mitnick story. Hafner, who was primarily responsible for the characterization of Mitnick as a "darkside hacker," admitted to Charles Platt for his 1995 review of *Takedown* that it "might have been a mistake to call him a darkside hacker." Hafner, in fact, has come to regret the characterization and what has followed from it. "There are malicious characters out there," she told Platt, "but Kevin is not one of them.... He has been turned into this bankable commodity. Leave the guy alone! He's had a really tragic life."[42]

That mistake has had a profound and lasting effect on Mitnick's life. Unlike hackers who seek publicity and visibility, Mitnick has always sought to maintain a low profile, even refusing to talk with Hafner and Markoff while they were writing *Cyberpunk*. As a result, Hafner and Markoff relied extensively on sources who portrayed Mitnick as a malicious, petty, and evil person who tampered with celebrities' telephone lines, altered credit reports, and accessed and changed police files, accusations that Mitnick denies. Two of the main sources for Hafner and Markoff's account were "Susan" and "Roscoe," two of Mitnick's fellow hackers who, as Hafner and Markoff write, "cooperated with us in the understanding that their true names would not be revealed." In a final touch of irony, the authors end the book with the line, "We respect their right to privacy." One of the two, "Roscoe," would later claim that much of the information he provided to Hafner and Markoff was intended to deceive them.

The most damning accusations against Mitnick were not his hacking exploits. What colored perception of Mitnick most thoroughly were the little things, most of which, Mitnick claims, were untrue and used for the purpose of "spin" and to "assign motive" to his actions. The accusation that seems to bother him most is the claim that he stole money from his mother's purse to further his hacking exploits, an incident that he refers to as absolute "fiction."

What has damned Mitnick in the eyes of both the public and law enforcement is not his hacking, but his personality. That characterization of Mitnick is built almost entirely on secondhand accounts from people who had either served as informants against him or had an investment in vilifying him to suit their own agendas. Turning Mitnick into the "archetypal 'dark-side' computer hacker" is a move that has suited a number of agendas, most recently Shimomura's and Markoff's.

Since *Cyberpunk*, Markoff has kept the Mitnick story alive in the pages of the *New York Times*, referring to Mitnick as "Cyberspace's Most Wanted," "a computer programmer run amok," and the "Prince of Hackers." Markoff also covered the break-in of Shimomura's system, which spurred the manhunt that would ultimately lead to Mitnick's arrest. Initially, the two stories were unrelated, the first describing how Mitnick was eluding an FBI manhunt (July 4, 1994) and the second detailing how Shimomura's computer system had been breached (January 28, 1995). Two weeks after reporting the break-in, Markoff was reporting that federal authorities had suspected that the "31-year-old computer outlaw Kevin D. Mitnick is the person behind a recent spree of break-ins to hundreds of corporate, university and personal computers on the global Internet" (February 16, 1995). From that point on, Markoff began telling the tale of the noble samurai warrior, Shimomura, versus the "dark-side" hacker, Mitnick — a true battle of good versus evil with an ending that seemed made-for-Hollywood. As Markoff concluded in his February 19 article for the *New York Times,* "Mr. Mitnick is not a hacker in the original sense of the word. Mr. Shimomura is. And when their worlds collided, it was obvious which one of them had to win."

The story, which gained considerable attention through Markoff's reporting, turned Mitnick's manhunt into a national event. It also resulted in the publication of three books about Mitnick and, ultimately, a feature film, completing the cycle. The story that left Los Angeles in print returned on the silver screen.

In the film version, Kevin Mitnick is offered up for sacrifice in a tale of good and evil that promises to further enrich both Markoff and Shimomura (they were reportedly paid $750,000 for their book deal, and one can only assume the movie option pushes them well over the $1,000,000 mark) and to completely demonize Kevin Mitnick in the public's eyes.

In July of 1998 Miramax announced that Skeet Ulrich would play the part of Kevin Mitnick in the film version of John Markoff and Tsutomu Shimomura's book, *Takedown.* The book, which chronicles the tracking and arrest of Kevin Mitnick, is the latest in a series of portrayals of Mitnick over which he has had no control. It is also, according to Mitnick, wildly inaccurate and libelous.

Scenes from the original film script included Mitnick whistling touch-tones into a phone receiver in order to make free phone calls (a technical and physical impossibility) and, most unbelievably, a scene in which Mitnick physically assaults Shimomura with a metal garbage can, leaving him "dazed, [with] blood flowing freely from a gash above his ear." The only difficulty with that part of the narrative is that Shimomura and Mitnick had never met, much less had a physical altercation, at that point in the story.

Mitnick remained a Los Angeles story. He was imprisoned in the Los Angeles Metropolitan Detention Center as a "pretrial detainee" for four years awaiting trial. During his detention, stories circulated about Mitnick that raised concerns so grave that Judge Marianne Pfaelzer went as far as denying him the right to a bail hearing. For those four years, Mitnick was held in a maximum-security facility, permitted visits only from his attorneys and his immediate family. His only contact with the outside world was on the telephone. Government attorneys refused to provide evidence to be used against him, citing its "proprietary nature," and an attorney of Mitnick's (a court-appointed panel attorney) was denied his fees (billed at the rate of sixty dollars per hour) by the court over the summer of 1997 because the judge ruled them excessive. Pfaelzer told attorney Don Randolph, "You are spending too much time on this case."

In an earlier case, Pfaelzer had prohibited Mitnick from unsupervised access to telephones while awaiting trial. The Office of Prisons found that the only way it could comply with the judge's order was to keep Mitnick separated from the general population. As a result, Mitnick spent eight months awaiting trial in solitary confinement.

In many senses, Kevin Mitnick can be seen as a creation of the press and the other media. The images they have generated have attached themselves to Mitnick's body and have had real and material effects. Most specifically, the denial of a bail hearing, which left Mitnick incarcerated for a period of more than four years awaiting trial, and his incarceration in solitary confinement both speak to the kinds of effects that such representations have had on his life.

While the media portrayal of Kevin Mitnick focuses on fears of the "evil genius" or lone computer hacker, there is a second story and hunt that focus on youth culture, rebellion, and fears of youth violence, particularly in urban settings.

Hacker Wars: LOD vs. MOD

While Mitnick was often framed as a daring thief or misguided genius, a second, more disturbing, set of descriptions characterizes hackers as deviant criminals, often violent in nature, through terms that range from the benign and comical to the serious and disturbing. In one of the more popular descriptions, hackers are labeled

as "dark-side" criminals, recalling Darth Vader and other mysterious powers of evil. In a less fantastic version they are simply
called "hoods," "menaces," or "cybervandals." Occasionally, however, these descriptions slide into more disturbing characterizations,
ranging from "twisted" to "serial hacker," with all the connotations
that are usually attached to the prefix "serial" (for example, serial
killer, serial rapist). The suggestion in each case is that hacking constitutes a kind of violent crime or is in some manner considered
deviant in the extreme.

A common characterization of hackers is to portray them as
violent criminals by connecting their typical youth rebellion to
more serious and more extreme forms of subcultural expression. A
"hacker war" between two rival groups, the Legion of Doom (LOD)
and the Masters of Deception (MOD), gained national attention
when the story was framed as a "gang war" fueled by issues of race
and urban and suburban unrest.[43] This hacker battle royal gained
national attention when this "boy war" erupted into headlines.

Initially, these hackers, even as they engaged in petty vandalism,
pranks, and boy games, were practicing a time-honored tradition
of youth culture and rebellion. One of the most basic means by
which boys express affection for one another is through a kind of
"affection through mayhem," the idea that a measure of violence between friends is actually a signal of their affection for one another,
rather than their animosity.[44] Things took a dramatic turn, however,
when members of the LOD found that they were beginning to move
beyond the boundaries of boy culture.

The "war," which reports played up by tapping into the cultural anxieties about gang warfare and urban youths run amok,
was in many ways the perfect expression of boy culture. As Anthony Rotundo has noted, "In their cultural world, where gestures of
tenderness were forbidden, physical combat allowed them moments
of touch and bouts of intense embrace. By a certain 'boy logic,' it
made sense to pay their affections in the coin of physical combat
that served as the social currency of boys' world."[45] For hackers,
the coin of social currency has shifted from physical combat to electronic warfare, and the two groups at the heart of the conflict were
expert and experienced enough at hacking to engage in full-scale as-

sault. The contest, which began as an eruption of tensions between two aspects of boy culture, would become something quite different by the conflict's end. As members of the LOD began to outgrow boy culture and needed to make the transition to the adult world, they found their rivals in the MOD doing everything in their power to make that transition difficult.

Both the LOD and the MOD have written histories that document the groups' origins, developments, active membership, and past or "retired" members. In these documents, each group presents its own historical narrative, mainly, as "The History of LOD/H" indicates, "to present an accurate picture of events and people who have been associated with this group."[46] The LOD text also illustrates how squarely the group fit in the mold of boy culture. The group's name (and most probably the idea for the group itself) originated with Lex Luthor. Chris Goggans describes the creation of the group: "The person whose idea it was to start the group, his handle was Lex Luthor, and from the DC comics, Lex Luthor's infamous group of antiheroes was The Legion of Doom, so it was a pretty natural choice. A lot of stuff has been attributed to it lately, such as it being a sinister type name. Well, Lex Luthor couldn't possibly have called his group anything other than the Legion of Doom. Anybody who has ever read a Super Friends comic knows that's exactly what it was called."[47] Luthor confirms the name's origin in "The History of LOD/H." He writes, "The name Legion of Doom obviously came from the cartoon series which pitted them against the Super Friends. I suppose other group names have come from stranger sources. My handle, Lex Luthor, came from the movie Superman I. In the cartoon series, LOD is led by Lex Luthor and thus, the name was rather fitting."[48] As the LOD grew, so did its reputation.

Although there have been relatively few actual members of the group, many hackers have claimed either to be members or to be in some way affiliated with the group in order to boost their own reputations in the underground. According to Goggans, throughout the late 1980s and early 1990s, the "reps of everybody involved in the group kind of sky rocketed" due mainly to the fact that, as a group, they all worked together and therefore "had a better resource of knowledge" than many of the other hackers working alone.[49] By 1986, the LOD had already established itself as "one of the oldest and most knowledgeable of all groups" and was credited in *Phrack* with having "written many extensive g-philes about various topics."[50]

As of 1990, the LOD documented only a total of thirty-eight members in the six years (1984–1990) of its existence, most of whom were marked as no longer active for reasons ranging from "No time/college" to "Lost interest" to "Busted" to "Misc. trouble." As of 1990 only six members were listed as "Current Members." By 1994, the group had officially disbanded.

Other than its notoriety, there was nothing particularly unique about the LOD. In fact, the LOD served as a kind of barometer for the hacker underground throughout the 1980s and 1990s. The group itself went through a series of transformations as new members joined and older members "retired." The LOD, according to Goggans, mirrored the underground, going through several boom and bust periods — the LOD "went through three different waves. You can kind of chart the history of the computer underground, it sort of runs parallel to the history of The Legion of Doom, because you can see as the new members came in, that's when all the busts happened. People would either get nervous about busts and move on and go to college and try to get a life, or they would be involved in some of the busts and some of them would end [and] leave that way. So it kind of went through three different membership reorganizations. You can tell who came in where because of what was going on. It finally kind of folded."[51] In essence, the LOD set the standard for computer groups nationwide and had, perhaps, the strongest reputation and greatest longevity of any group in the underground. The group recruited from the best nationwide, with members ranging from California to New Jersey, from Georgia to Minnesota — spanning fifteen states in all and one member from the United Kingdom. The end of LOD was marked by the formation of a computer security group, COMSEC, which was several of the LOD members' effort to bridge their hacking talents into the world of adult responsibility. It was also the site of tension that would ultimately lead to the arrest of a number of MOD hackers and the end of the hacker war.

In distinction to the LOD, the MOD was formed later, in 1987, was a much smaller group, and was locally based in New York. The three original members were Acid Phreak, the Wing, and Scorpion. Later Phiber Optik would join the group, and by August of 1990, the MOD had grown to fourteen members. The MOD shared few characteristics with the LOD, either geographically or experientially, and the differences between the two cultures would be the basis for the hacker war that followed.

The most interesting, and certainly most detailed, account of the rivalry between the LOD and the MOD is that in Michelle Slatalla and Joshua Quittner's book, *Masters of Deception: The Gang That Ruled Cyberspace*. Slatalla and Quittner put forth a narrative driven by racial conflict that portrays the rivalry between the LOD and the MOD as a "gang war." As Slatalla and Quittner describe it, the conflict between the LOD and the MOD was motivated by two main factors. First, there was an incident over a phone "conference bridge" (an event where a group of hackers gather on phone lines and do the equivalent of a conference call) in which an LOD member called an MOD member (John Lee) a "nigger." This, Slatalla and Quittner contend, changed a "friendly rivalry" into an "all-out gang war," a "highly illegal battle royal."[52] The second factor, which seems particularly odd for a discussion of cyberspace, was, according to Slatalla and Quittner, geographical. In fact, Slatalla and Quittner subsume the first event within the second. As background to this second point, it should be noted that the LOD had members from Texas. They were on the line during the crucial conference bridge when John Lee of the MOD (as this point he was using the handle "Dope Fiend from MOD") joined the conversation. One of the Texans noted over the phone that the newcomer did "not have an accent common to these [that is, Texas] parts." The newcomer spoke in a "distinctly non-white, non-middle-class, *non-Texas* inflection." As Slatalla and Quittner report the event, "One of the Texans (who knows who) takes umbrage. 'Get that nigger off the line!' "[53]

At that moment, as the narrative unfolds, John Lee (an African American) decided that he would take revenge on the LOD — "With that one word, war had been declared." As Slatalla and Quittner argue, "You don't survive on the street by allowing white boys to

call you nigger." John Lee retaliated by spreading "secret" LOD
files around the hacker community. But, curiously, the racialized,
class narrative gives way to the geographical one — "The battle lines
were drawn now.... Scott and Chris [two of the Texans in the LOD]
didn't care anymore about the specifics than John did the day he
heard someone calling him nigger over a conference bridge. *It was
New York City against Texas.*"[54]

Despite hackers' insistence that race has never been an adequate
criterion for judging people, the boy culture element of the hacker
underground has always been (and continues to be) rife with racism.
But Goggans, one of the best known and most influential members of
the LOD, has distilled Slatalla and Quittner's book with disdain: "So
we end up with 'New York vs. Texas! Black vs. White!' Forget about
the fact that LOD had a black member and a Mexican member,
forget anything like that. It's 'Three racist redneck Texans against a
multi-ethnic group of computer youths from the inner city of New
York, working to improve the lot life handed them!'"[55]

Following the publication of an advance chapter of the book in
Wired, Goggans responded by repudiating the entire narrative, calling "the whole racial issue" a "non-issue" and by claiming that "to
imply that such things were strictly New York-versus-Texas issues is
ludicrous." According to Goggans, who saw the event unfold, it was
a case of boy culture clear and simple — battles for control, technological dominance, and mastery were the issues at stake. Race, he
argued, was just a way to make the story more sellable. Goggans's
explanation of how and why rivalries exist and are created in the
underground documents the manner in which struggles for independence and control often erupt into conflict: "There have always been
ego flares and group rivalries in the underground; there always will
be. The Legion of Doom was founded because of a spat between its
founder (Lex Luthor) and members of a group called The Knights of
the Shadow. These rivalries keep things interesting and the community moving forward as hackers butt heads in an attempt to find the
newest bit of information in healthy one-upmanships. MOD was
different. It took things too far. And not just against two people in
Texas."[56]

Part of what had gone too far was the MOD's disruption of former LOD members' computer security business. Just as Goggans and others were moving out of their boy culture and into the world of corporate responsibility, they would find MOD members making their lives difficult. The boy war had escalated into something beyond hacker one-upmanship. Goggans and other LOD members were trying to make a living, while MOD members were still waging warfare as youths struggling for control. At that point, LOD members did the unthinkable: they called the authorities.

The narrative construction of the conflict between the LOD and the MOD as a matter of "gang warfare" serves to define hackers in terms of corporeality and physical space. Slatalla and Quittner's characterization reconnects the concepts of hacking and the body through race and the threat of violence. Equally important are the connections to geography, highlighted through the narrative of class — the MOD's urban, street-smart hackers, versus the privileged suburbanite kids from Texas. What Slatalla and Quittner miss, however, is the shift that was occurring in the hacker underground. The boys of the LOD were becoming adults, and making that transition meant entering the world of adult responsibility and adult authority.

The indictment that would lead to the sentencing of the members of the MOD, however, was completely unrelated to any sense of corporeality or geography. It contained eleven counts, which could result in up to fifty-five years in jail and almost three million dollars in fines. The formal indictment charged "unauthorized access to computers, possession of unauthorized access devices, four counts of interception of electronic communications, and four counts of wire fraud."[57]

The most interesting charge leveled against the hackers was one of conspiracy. It was alleged that the members of the MOD had conspired to "gain access to and control of computer systems in order to enhance their image and prestige among computer hackers."[58] For the first time, hackers were accused of organized crime. By elevating the indictment to include conspiracy, prosecutors had criminalized one of the oldest and most basic components of the computer underground — the desire to build and maintain a reputation based on group affiliation. The effect of the indictment was to suggest that the hackers of the MOD were in fact *violent* criminals who were putting the nation's information infrastructure at risk. Their associations were like those of other organized crime groups, the mafia or gangs, for example. The suggestion was that the MOD was capable of violence, if not physically directed at the body, then a kind of virtual violence, the equivalent of brute force in virtual space.

Such characterizations position hackers in terms of corporeality — not in terms of a loss of physical presence but as a threat to the physical world. Existing in a virtual space, which is nonetheless connected to the physical world, hackers are capable of a new breed of violence, directed at the body from nonphysical space. For law enforcement and the media, troping on the increasing fears of street-gang violence and organized crime, these hackers are represented with the same terms as John Gotti or Crips and Bloods. What began as a boy war had entered the arena of adult responsibility. But the story that grabbed the headlines had little to do with the transition that was occurring. Instead, headlines focused on the threat that hackers posed and attempted to link hacker activities to everything from race riots and gang warfare to organized crime. Just as the members of the LOD had a difficult time breaking free from the boy culture that they had a hand in shaping, the popular press was reluctant to let them make the transition smoothly.

The hacker war set the stage for a new set of anxieties that would be played out throughout the 1990s. Just as fears of urban unrest and marauding youths had been used to characterize the MOD hackers, other hackers would find themselves the object of similar displacements of cultural anxieties. The charge of conspiracy, used to prosecute the MOD, would also find its way into the next case, but the sense of organization would become even more refined and, ultimately, directed toward issues of national security.

Kevin Poulsen

The third type of characterization of hackers carries the suggestion that hackers are by the very act of hacking violating national security. In these narrative constructions, hackers are seen as "notorious," as "rogues," as "Most Wanted," as "invaders" and "intruders," even as "computer terrorists." In contrast to depictions of hackers as "master criminals" or even "gang members," these constructions are, perhaps, the most serious and the most exaggerated of the three.

In many ways the most interesting narrative of the pursuit and capture of a hacker is that involving Kevin Poulsen. Poulsen, who used the handle Dark Dante, was accused of, among other things, intercepting Pacific Bell security conversations and embezzlement of government information, including stealing a computer printout that contained information about how phone numbers were assigned (rendering them vulnerable to phone taps) and even the phone numbers of celebrities and leaders such as Ferdinand Marcos. A number of his discoveries, it would turn out, were targets of highly confidential FBI investigations, and, as a result of this and other offenses, Poulsen would be charged, ultimately, with espionage.[59] Although most of the more serious charges (including espionage and most of the wiretap charges) were dismissed, Poulsen was still convicted of a host of charges and sentenced to five years in jail.

Like most hackers, Poulsen's interest in computer networks started as a matter of curiosity, but quickly turned into something more. "My intrusions," he explains, "particularly physical ones, were more than just ways of gaining knowledge. I think, in a way, part of me saw the network as something mystical and arcane. Exploring a telephone switching center, immersed in the sights and sounds of rooms full of equipment, was a kind of transcendence for me. A chance to become something greater than myself."[60] As a consequence of his hacking, Poulsen was fired from his job in Northern California and moved back to Los Angles in 1988. Unemployed, he continued hacking as a means to generate income, even developing an elaborate scam to utilize disconnected escort-service phone numbers to supply Los Angeles pimps with a steady supply of customers. When escort services would go out of business (usually as a result of police raids), their numbers would be disconnected. What Poulsen figured out was that their advertising, particularly in the Yellow Pages, meant that customers were still calling. It was merely a matter of reconnecting and redirecting the calls for services that were already advertised and marketed.

Some of Poulsen's exploits had received the attention of federal investigators, however. After discovering he was under investigation, Poulsen hacked into the FBI's systems and discovered a maze of wiretaps and surveillance programs that were monitoring everyone and everything from the restaurant across the street from him to (allegedly) Ferdinand Marcos. As Poulsen explains it, "I became concerned that I, my friends, or my family, might be subject to surveillance by either Pacific Bell's security department or the FBI. This concern prompted me to actively research the physical and electronic surveillance methods used by these agencies. It was while conducting this research that I stumbled (via my computer) across the FBI's wiretap of a local restaurant."[61]

The crimes that Poulsen was convicted of centered on what happened after his return to Los Angeles. Specifically, as Poulsen describes it, "On a more practical level, the knowledge I gained of the phone network, and my access to Pacific Bell computers, allowed me to increase my odds of winning radio stations' phone-in contests substantially. I played these contests in part because it was a challenge and it allowed me to engage in the sort of complex, coordinated efforts that I missed since the loss of my career. I also saw it, at the time, as a 'victimless' way of making money with my access."[62] After all was said and done, Poulsen had won two Porsches, a trip to Hawaii, and tens of thousands of dollars in prize money. But these crimes were not the one's that led to Poulsen's arrest.

On April 11, 1991, Kevin Poulsen was leaving Hugh's Market in Van Nuys, California, near midnight when he was apprehended by a bag boy from the market who had recognized Poulsen from an airing of an *Unsolved Mysteries* episode. According to Jonathan Littman, the bag boy, who chased, tackled, and held Poulsen until authorities arrived, told the authorities "that they could have their suspect now."[63] The *Unsolved Mysteries* episode, which was in large part responsible for Poulsen's apprehension, focused primarily on misleading claims that Poulsen broke into secret government computers and posed a serious threat to national security. (One of the charges against him, which was later dismissed, was for "gathering of Defense Information" that had been "classified 'Secret.'")[64]

What is most intriguing about Poulsen's case is the sense in which he is characterized as perverted and dangerous. As Littman's book title testifies, Poulsen is regarded as "twisted" and as a "serial hacker."[65] To law enforcement, however, what makes Poulsen dangerous is not the secrets he knows but, rather, *the kinds of secrets he knows.* In many ways, the information that Poulsen uncovered in his surveillance of the FBI was incidental. The disturbing thing about Poulsen was the knowledge he had gained about the *kinds* of people that were under surveillance. Poulsen exposed not the secret per se but the secret that guards all secrets. In a panoptical environment, the power of the gaze is defined by its possibility, by the fact that it is always possible that at any given time one could be seen. Poulsen's discovery, and therefore his threat, was the ability to know, at any given moment, who was and was not being watched. Poulsen's threat was not to any particular secret but to the very structure of secrecy itself.

The need to brand Poulsen as a "threat to national security" was based, at least in part, on his ability to elide surveillance, and what that ability reveals is the degree to which surveillance defines the state of "national security" in the digital age. Ironically, the government turned to *Unsolved Mysteries,* revealing the manner in which such shows are complicit with strategies of government surveillance in part through encouraging citizen participation.

Poulsen's response to the show enacted an intervention into the mechanisms of surveillance. Poulsen had been informed of the time and date that the show would run and realized that the attention it would bring would potentially lead to his arrest. In response, he formulated a plan to short-circuit the system. Tempted initially to "knock out Channel Four," by cutting cables at the transmitter tower to block the airing of the show in Los Angeles, Poulsen reconsidered, realizing that it would "guarantee a repeat appearance on every segment of *Unsolved Mysteries.*"[66] Instead, Poulsen's response was more creative. Littman reports the following about the night of the show's Los Angeles airing:

> On schedule, NBC plays the show's eerie theme music followed by a quick preview of that night's episodes. . . . Then, in a matter of seconds, everything changes. "I'm dead!" calls out an operator, peeling off her headset. "Me, too!" another cries, and then like an angry flock of blue jays the voices squawk. "I'm dead!

I'm dead! I'm dead!" Rajter looks at his watch: 5:10 p.m. Every phone line in the thirty-operator telecommunications center is dead.[67]

Rather than cutting the outgoing transmission, Poulsen had entered the phone switch on his computer and disconnected the eight hundred lines that fed into the show's response center. The show would go out as scheduled, but viewers were unable to call in their tips. Poulsen would nonetheless be characterized on the show as engaging in espionage, and it would be that broadcast that would ultimately lead to his arrest. In this case, the stakes had been raised significantly. Even though the charges of espionage were dropped, Poulsen (and hackers everywhere by extension) had been branded as a threat to national security, playing on the worst technophobic impulses and anxieties of contemporary culture.

Hackers and the Displacement of Anxiety

These three chronicles of hackers and hacker activity clarify precisely how anxiety is displaced from broader social issues onto the figure of the hacker. Such accounts also cover over, in each case, the same issue — the development and deployment of surveillance technology by law enforcement agents and the extreme lengths that they have gone to and will go to in order to apprehend hackers. The narratives that emerge are twofold. First, each narrative effects a certain displacement, whereby the hacker is made to stand in for an issue of great cultural anxiety — the loss of the body and identity; the threat of violence; and the fear of threats to national security. Second, these narratives of anxiety provide a justification for the increased deployment of technology aimed both at creating a panoptical virtual space and at depriving those in that virtual space of the secret that connects their virtual identity to their physical bodies.

The discourse about hackers and hacking is related to a broader social discourse concerning the contemporary relationship between technology and society. That relationship is manifested as anxiety when society, generally, is uncertain or unclear about the implications of the technical for its way of life or its well-being. Whether it is fear of violence, race, or class, as was the case with the MOD, or the threat to national security and secrecy, as was the case with Kevin Poulsen, hackers are continually branded as criminals in remarkably flexible and varied ways. While this anxiety is recorded throughout the greater part of human history, it reached a certain peak in the latter half of the twentieth century, particularly in response to what we might consider the proliferation of computer technology brought about by the "home computer" or "personal computer" (PC) as computers have entered the home, workplace, and schools.

Anxiety manifests itself around the notion of an "expectation," whereby we find, according to Freud, a "general apprehensiveness, a kind of freely floating anxiety which is ready to attach itself to any idea that is in any way suitable, which influences judgment, selects what is to be expected, and lies in wait for any opportunity that will allow it to justify itself." In other words, anxiety is related to a sense of the unknown and of uncertainty. This particular form of anxiety — what Freud called "expectant anxiety" — clearly manifests itself around the notion of technology. Such anxiety is different from what we commonly think of as a "phobia." With computer technology, people are not, necessarily, afraid of the machines themselves. What they fear is the future — as Freud argues, they "foresee the most frightful of all possibilities, interpret every chance event as a premonition of evil and exploit every uncertainty in a bad sense."[68]

The anxiety over technology, as an expectant anxiety aimed at the future, calls into question almost every aspect of daily human interaction. Accordingly, such anxiety triggers the process of displacement, whereby the expectant anxiety over something important can be rethought and managed in relation to something unimportant. Computers, being both part of everyday experience and a ubiquitous part of daily life, are the idea vessel for such displacement. The process of displacement occurs through allusion. In such an act, the object onto which anxiety is displaced is "easily intelligible" (unlike allusions in dreams, for example), and the "substitute must be related in its subject-matter to the genuine thing it stands for."[69] Technology, in most every sense, is not the cause of these fears but rather the object tied to anxiety by allusion or proximity. In the discourse surrounding hackers and hacking, the main site of displacement is the body itself in response to the challenge that technology poses to corporeality and the structure of the law, specifically, and the structure of society, more generally.

This sense of anxiety, attached to the physical body, is an anxiety that precedes the computer. The fear attached to the body that begins with the discourse of television reaches its zenith with discussion of "virtual reality" and the actual disappearance of the body. This anxiety can be located in the movement from the physical to the virtual. The threat to the body had already been well rehearsed in connection to television viewing for two decades. Television, like most forms of technically mediated communication, can be considered a virtual medium. In television's early development, a great deal of anxiety was centered on the fear that the medium would either sap one's strength, addle one's brain, and decimate one's critical reasoning abilities, turning one into a "couch potato," or produce a generation of juvenile delinquents. In short, the television was able to alter or disable the physical through a participation in the virtual. The introduction of the video game, and later the PC, met similar criticisms. The virtual came to represent, as a manifestation of anxiety about the technological, a threat to the body.

The latest step in the transformation into virtual reality threatens the body at the most basic level — the virtual threatens to make the body disappear. But it also threatens the body at a different level. This anxiety is an outgrowth of the first — the threat that the virtual poses to the body itself. If the first case, the virtual, is marked by a disappearance of the body, the second case, a return to the physical, is marked by a fear of the virtual. In other words, the threat that the virtual poses to the physical is akin to the threat a ghost or specter poses to the living. A virtual presence is a threat to the living precisely in terms of its incorporeal existence. The virtual "haunts" the physical world. It is a dead presence, a "spirit."

The regulation of the hacker is accomplished by juridically creating a corporeal subject who can be monitored. As a result, criminality is defined solely in terms of the performance of an identity and the ability or inability to match that identity to a corporeal presence. Like ghosts that haunt and terrorize the living, hackers who maintain a virtual existence remain beyond the grasp of the law and in doing so are able to elude the gaze of state surveillance. The criminalization of virtual identity becomes both the goal of law enforcement and the primary locus of discussion about the threat of hackers and cyberspace to society. As law enforcement describes it, the loss of the body to virtual identity not only serves to make the body of the hacker disappear but also makes the hacker legally *unaccountable,* providing a space from which the hacker may then inflict violence on bodies (or even on the social body, as with the threat of espionage) with impunity.

As the cases of these hackers make clear, there is an investment in creating stories and images of hackers that serve to allow a greater social displacement of anxiety upon them. As part of that system of representation, law enforcement, media, and the state are heavily invested in the manner and style in which hackers are represented. Those representations, whether translated into law or broadcast on *Unsolved Mysteries,* are useful vessels for the displacement of anxiety about crime and technology and, ultimately, provide a powerful diversion to allow the progression of state surveillance, observation, and regulation and to include and encourage citizen involvement in the processes and mechanisms of surveillance.

Epilogue

Kevin Mitnick and Chris Lamprecht

From the hackers of the 1950s to the present, secrecy has played a crucial role in understanding the place of technology in contemporary culture. It has also been at the center of understanding who hackers are and what they do. Two recent cases have served to illustrate the merger of secrecy, criminality, and the displacement of anxiety onto the figure of the hacker: first, the case of Kevin Mitnick, discussed in chapter 6; and, second, the case of Chris Lamprecht, a hacker who was convicted of theft but sentenced for being a hacker.[1]

These two cases provide insight into the degree to which corporate secrets, media and popular culture representations, and the displacement of anxiety can combine to have real and profound effects on the lives of hackers caught in their wake. While it is clear in each case that these hackers have committed crimes, what is interesting about their cases are the ways in which their position as hackers has been used to justify what seem to be extreme measures against them, government and court responses that are far out of proportion to their crimes.

Technology and Punishment Redux: Kevin Mitnick, 1999–2000

As described earlier, in February of 1995, Kevin Mitnick was arrested following one of the most intense hacker manhunts in history. Since that time, he has been tried in the state of North Carolina, faced a twenty-five-count federal indictment, and served four years as a "pretrial detainee," eventually reaching a plea bargain with government attorneys. Although Mitnick gained no profit from his alleged hacks, he faced charges of both computer and wire fraud that could

Protesters outside Miramax's New York office. Source: *2600* Magazine.

have resulted in more than twenty years in prison. Stories, including half a dozen books and a film set to be released by Miramax, about Mitnick's case have all focused on his capture and arrest. What has been less publicized has been the aftermath of Mitnick's case, which has included a number of important legal matters and has spawned a unified hacker movement that has focused on the political issues surrounding his incarceration. A generation of hackers, some in their early teens, were the force behind the "Free Kevin" movement, which held protests, engaged in a public awareness campaign, and even created a legal defense fund for Mitnick. While Mitnick's hacking may have had an impact on the underground, his arrest and imprisonment shaped the attitudes and opinions of an entire generation of hackers who came of age in this five-year period.

What Mitnick's case reveals, more than anything, is the manner in which secrecy is tied to technology. The most serious charges against Mitnick involved the copying of proprietary information from cellular phone companies. The violation was not breaking into the system, nor was it the actual copying or possession of files from the system. Instead, what was violated was the proprietary nature of those files. Mitnick broke the code of secrecy, which, according to the corporations in the indictment, had made the information worthless.

Mitnick's case raised a number of important issues that were hotly debated in pretrial motions but never litigated. Four central issues leading up to the plea bargain made his case unique and reveal a great deal about legal and public attitudes toward hackers. First, Mitnick's pretrial detention and lack of a bail hearing were used by Mitnick's supporters as a fundamental rallying point. Second, advocates of the "Free Kevin" movement argued that the government's effort to deny Mitnick and his attorneys access to the evidence to be presented at trial was a violation of his fundamental rights. Third, there were a number of moments of resistance where hackers sent messages to the larger computer community, the most visible being a hack of the *New York Times* Web page, done the weekend the Starr report (on dealings of President Clinton) was released. Finally, the plea bargain itself, which came in the wake of several defense motions accusing the government of illegally gaining evidence against Mitnick, drew the case to a close, but not without a media feeding frenzy, which included serious leaks to the media.

Pretrial Detention and Bail

As noted earlier, prior to his plea agreement, Mitnick spent four years at the Los Angeles Metropolitan Detention Center. The MDC is a jail, rather than a prison, where most offenders in Los Angeles are housed before trial. The length of time spent at the MDC ranges from only a few days to a year. Mitnick's attorneys argued that his pretrial incarceration, which lasted over four years, was punitive, something prohibited by law. During that time, Mitnick was not allowed visitors other than his attorneys and family members, and he was not permitted to work or have many of the privileges that convicts are allowed.

Mitnick's pretrial incarceration resulted from his lack of bail—he was never given the opportunity to have a bail hearing. Judge Marianne Pfaelzer denied the request for bail without hearing, meaning that the government never had to show cause or justify why Mitnick should remain in jail while he awaited trial. The denial of a hearing was appealed up to the Supreme Court, which refused to hear the case. This would be the first "outrage" that would begin to unify the "Free Kevin" movement under the principle that he was denied his constitutional right to a bail hearing. From that point on, the focus would be not on his guilt or innocence but on the process by which his case proceeded. It was also the moment at which Mitnick's own attitude in his case shifted and he began to describe himself as a "political prisoner." Mitnick and others close to his case were convinced that he was being treated harshly to send a message to other would-be hackers. In the months that followed, the court and the government would face new legal issues, many of which had previously been undecided, making it exceedingly difficult for Mitnick and the defense team to gain access to the information that would be presented against Mitnick at trial.

Kevin's Computer

After Mitnick's arrest in 1995, the government was in possession of two of Kevin's laptop computers, containing thousands of files and nearly ten gigabytes of data. When attorneys for Mitnick and his codefendant, Lewis DePayne, asked to review the evidence, they were provided with a 187-page list of filenames with no summary or explanatory information — defense attorney Richard Sherman described the data as "incomprehensible." Although the government claimed to have examined each piece of data, it claimed no record of what is actually on the computer hard drives. The investigators, the government claimed, who sorted through the thousands of files took no notes and made no record of what they found.

The question of access to a computer was only surprising insofar as it is an issue at all. Nearly all of the evidence in this case is in electronic form, and it is patently clear that Mitnick's expertise was undoubtedly useful, if not essential, in preparing his own defense. With roughly five million pages of material (a large percentage of it, 70 to 80 percent according to prosecutors, containing source code or software programs), electronic searching and retrieval were extremely useful in cataloging and analyzing the information. These were Mitnick's files, after all, and he was probably the best person to decipher what is on his own hard disks.

What causes concern is that the computer was being targeted as such a threatening device. While it is true that Mitnick used a computer to commit crimes, it is hard to imagine a judge prohibiting someone accused of forgery from using a pen, or a bookie from making phone calls, even if that is how she or he took bets. Without a modem or network card, the threat a computer poses is minimal to nonexistent. Perhaps prosecutors were wary because they didn't understand that, or perhaps they were being punitive, knowing what an important part computers have played in Mitnick's life. Perhaps (as one report has it) they were afraid Mitnick would spend time playing computer games, rather than preparing his defense. In any event, it was not the first time that Mitnick had been put in a difficult position because of his technological savvy. Earlier, he had spent eight months in solitary confinement as a way to prohibit him from using the telephone, out of fear that he would whistle a computer virus over the phone lines. Later, while in jail, he was again sent to solitary confinement and all of his items were confiscated after it was suspected that he was modifying a Sony Walkman to create a transmitter or electronic eavesdropping device. Both these imagined tricks were technologically impossible; authorities' belief that they could be accomplished was prompted by an overreaction based on a fear of technology, generally, and a fear of hacking, specifically.

The data in this case was unusual and presented more than a few problems for both Mitnick and the court. What makes these problems unusual (a fact that Judge Pfaelzer steadfastly denied, claiming there was "nothing unusual" about the case at all) is the nature of the data itself. The important data fell into three categories. The first is the least interesting and, probably, the most damaging. Although the government gave little indication of any of the evidence that would be presented, it was clear that some of it would be composed of correspondence and text files that document illegal activity that was undertaken by the defendants. Files containing stolen passwords and credit card numbers, for example, would have been necessary for the government to make its case on most of the indictments. This data, which the government *must* have known about, should have clearly followed traditional rules of discovery, which is to say, it should have been provided to the defense just as any other documents or evidence must be. That, however, didn't happen until much later (three and a half years) in the case, and the reason for that was related to the second two categories of evidence.

The second category, what was labeled "proprietary software," consisted of the objects that Mitnick was accused of stealing. It was also the source of most of the damages for which Mitnick was being held accountable. This software, it is alleged in count 15 of the indictment, is valued in the millions of dollars and is the source of the massively inflated jail time (up to two hundred years) that Mitnick was faced with. Because he was charged with wire fraud (in addition to computer fraud), the sentence was based on the damage done, rather than the offense. Because he had "compromised proprietary software," the prosecutors charged, he made that software, very expensive software, worthless. What Mitnick had violated was its *proprietary* nature, and that, the government claimed, was what gave it its value. It did not matter that Mitnick did not deprive the company of the information or that there was no evidence that he would sell or distribute it. What mattered in this case was that the value of the software was generated by its secrecy, and by copying it, Mitnick had violated that secrecy, rendering it worthless. This proved a problem for the rules of evidence as well.

The government, however, argued that because this software was proprietary, it couldn't make copies to provide to the defense. Copying it, they said, would amount to a commission of the same crime for which Mitnick was under indictment. The companies involved, the government argued, were none too happy about having a second copy of their source code produced. The prosecutors offered to let the defense team look at the evidence at the government offices, but because it is proprietary they were unwilling to copy it. Of course, this presented a problem for Mitnick in particular; being held in the Metropolitan Detention Center made it impossible for him to visit the government law offices.

The third category of evidence presents an altogether new challenge — encryption. At least a portion of the files on Mitnick's hard disks was encrypted, and Mitnick was not forthcoming with the password. The government claimed that it had no way of knowing what was in those files, so it didn't plan to use them. Therefore, they argued, there was no need to provide them to the defense. Legally, this presented a problem. The defense claimed that some of that information could be exculpatory — that is, it might be very useful in Mitnick's defense. They wanted a copy of the encrypted files to find out. But if that was allowed, the prosecution, suspecting that some of the most damaging information was probably encrypted, wanted to look as well. This, the defense argued, violated Mitnick's right to not incriminate himself. The government had data it couldn't use, and what the defense might be able to use, it couldn't have. The issue was resolved by the judge based on an appeal made by the prosecution. Those files, the prosecution claimed, could contain a virus or damaging software, and that virus or software could be released by Mitnick if the files were returned to him. Judge Pfaelzer saw the encrypted files as a clever move on Mitnick's part and ultimately refused the defense team access to the files unless Mitnick revealed the password to decrypt the files.

The nature of the electronic data, for which there had been no case law or precedent, was used by the government as a justification for not providing copies of any of the electronic evidence to be used against Mitnick to the defense. The complexity that technology brought to the case was used by the government as a smoke screen to deny basic access to information about the case.

The government also used the technological nature of the evidence as a way to restrict access to it. In response to industry and government concerns about distributing proprietary software to the defense team and to Mitnick, in particular, both sides agreed to place the proprietary software under a protective order, meaning that the evidence could not be discussed outside of the defense team and their experts. Government attorneys, however, wanted to extend that protective order to cover all the evidence in the case, including the categories of "hacker tools" and Mitnick's personal correspondence.

A protective order that covered all of the evidence to be used in this trial would have prohibited Mitnick from discussing any aspect of the case publicly and would, in essence, have prohibited him from being able to tell his own side of the story at a later date. Even if he were found to be not guilty, the protective order would make it impossible for Mitnick to discuss the charges against him or the indictment without being held in contempt of court. As a result,

Mitnick has refused to sign the protective order, which he believed was an attempt to "chill free speech in the future."

As the initial hearing proceeded, other facts began to come to light. The defense claimed that one of the witnesses against Mitnick, Ron Austin, had been working as a government informant while he was employed by Mitnick's former attorney, Richard Sherman, and during that time was privy to privileged attorney/client communication. In 1994, the defense claimed, Austin "was surreptitiously (and apparently illegally) tape-recording conversations with Kevin Mitnick as part of his cooperation agreement with the government."

Although the government filings indicate that Austin's employment in Sherman's office was "unbeknownst to the government," Assistant U.S. Attorney David Schindler " 'authorized' Austin to continue to surreptitiously record his telephone conversations with Mitnick 'to facilitate the investigation' " in 1994. The defense argued in a filing that "government's failure to cease all interviews with Mr. Austin immediately upon the disclosure of his relationship with Mr. Sherman constitutes, in itself, a serious abrogation of the government's professional, ethical, and legal obligations." According to Mitnick's attorneys, the conditions of Ron Austin's plea bargain, which they requested in October of 1996, were not made available to them for nearly two years after the initial request. Such information, Don Randolph says, can be essential in impeaching witness testimony.

Shortly before Mitnick's plea deal was struck, the defense team introduced two motions to suppress evidence. In the motions, they argue, the search warrant that was used in Mitnick's arrest was so overly broad (it failed to include an address, physical description the building, or other identifying information) and was so badly executed that it should be considered invalid. More damaging, though, were allegations that Shimomura was acting as a government agent (confirmed in a statement by FBI director Louis Freeh) when he illegally monitored and intercepted several of Mitnick's communications. Since that was the grounds for "probable cause," the defense argued, anything resulting from those illegally intercepted communications must also be suppressed. If successful, this would suppress nearly all of the evidence against Mitnick, making a conviction extremely difficult. The motions were taken seriously enough by prosecutors to request a delay in having them heard so that they could formulate a response. That response would not be filed. Instead, that week, prosecutors and defense attorneys filed a plea agreement with the court that ended the case before a decision could be rendered on either motion.

While the legal issues presented new challenges both for the court and for the attorneys, the case also produced a range of responses in the hacker community, which ranged from indifference to outrage. In a few instances, hackers took matters into their own hands. There were protests, editorials, Web pages, and bumper stickers and T-shirts with the "Free Kevin" logo. The most dramatic incident was a high-profile hack, which took down the *New York Times* Web site for the better part of a day. While hacked Web pages are fairly commonplace, and occur with surprising regularity, they usually fall into the category of juvenile pranks, where no damage is done — Web page images are either replaced with pornographic ones or hackers leave messages to document their hacking talents. The *Times* web hack was different, primarily because it was done as a political hack.

Hacking the *Times:* What Makes This One Different?

On Sunday, September 13, 1999, the *New York Times* Web page was hacked by a group calling themselves "Hackers for Girlies." The hack forced the *Times* to take its site offline for nine hours, hours of what happened to be potentially one of the biggest days in the paper's online history. According to the *Times,* traffic, which was already up 35 percent on Saturday, was expected to double on Sunday, in large part due to the *New York Times* release and coverage of the Starr report. The hack, which was by all accounts a fairly sophisticated attack, was different from most previous hacks in a number of ways. Although there had been hacks of Web pages designed to express political dissatisfaction (for example, the Department of Justice Web page hack regarding the passage of the Communications Decency Act and hacks of both the Conservative and Labour Parties in Britain prior to elections), this was the first hacking incident that could be considered a *political intervention* — in this case, a major media outlet was financially damaged because of the public perception that its coverage was unfair.

Although much of the text of the hacked Web page centered on the hackers' dispute with Carolyn Meinel, author of *The Happy Hacker,* much of the criticism directed at Meinel can and should be seen as hacker infighting. What the page's comments reveal is that these hackers were concerned with the larger question of how they were being portrayed. "The real reason we put any blame on Carolyn Meinel," the hackers write, "is because of her obtuse over-dramatizations of our actions."

At the heart of the dispute was Mitnick's case, and at issue, for these hackers, was the question of John Markoff's coverage of Mitnick's case. As discussed earlier, Mitnick was the subject of several *New York Times* stories written by Markoff and ultimately the subject of a best-selling novel co-authored by Markoff and Tsutomu Shimomura, the San-Diego-based security expert who helped the FBI track and capture Mitnick.

Hackers alleged that Markoff used his position at the *Times* to "hype" the story of Mitnick's arrest and capture and to "demonize" Mitnick in the public imagination. These perceptions, according to the hacker community, accounted in large part for Mitnick's long incarceration in a maximum-security jail and his denial of the right to a bail hearing.

In short, the message that hackers left on the *New York Times* Web site could be boiled down to one simple fact: they felt that the way hackers are covered by the mainstream media generally, and the *New York Times* specifically, is unfair. They were disturbed both by what had been written about them and by what stories had been overlooked. The battle over such representations continued to be played out, ironically enough, in the coverage of the hack itself. Mitnick learned of the incident over a local Los Angeles news radio station, where he heard the hack described as an "act of Internet terrorism." Mitnick, who was, at the time, only four months from trial, was upset by both the incident and the subsequent coverage of it.

The message that the hackers left that Sunday came in two parts, the page that was displayed on the *Times*'s Web site and the comments left in the HTML code, which were far more articulate than the hacker-speak that appeared on the surface, as the hackers themselves indicate in their P.S.: "0UR C0MMENTS ARE M0RE 'LEET THAN 0UR TEXT. DOWNLOAD THE SOURCE T0 TH1S PAGE AND P0NDER 0UR W1ZD0M." The hackers described the comments embedded in the page's source code as "the real meaning" of the page, including supporting quotations from Tennyson, Voltaire, and Milton.

The hackers' central grievance stemmed from what they saw as Markoff's involvement in the pursuit and capture of Mitnick. The Web page's message targeted Markoff specifically, asking: "D0 YOU HAV3 N1GHTMAR3S ABOUT H3LP1NG 1MPRIS0N K3V1N? KN0WING THAT Y0UR LI3S AND D3C3IT H3LP3D BR1NG D0WN TH1S INJUST1C3?" What lies beneath the code in the comments spells out the hackers' complaint more directly: "The injustice Markoff has committed is criminal. He belongs in a jail rotting instead of Kevin Mitnick. Kevin is no dark side hacker. He is not malicious. He is not a demon. He did not abuse credit cards, distribute the software he found, or deny service to a single machine. Is that so hard to comprehend?"

Markoff denied that his coverage of Mitnick's case was anything other than objective. After years of covering Mitnick and because of his close connections with Shimomura, Markoff found himself with "access to remarkable events" that he says "I wrote about as accurately and clearly as I could." "There were no dilemmas," he said. "I told my *Times* editors what I was doing every step of the way." Regarding the decision to hype the story, Markoff responded, "I didn't place the story." If hackers were upset about how the story was hyped, Markoff thought they were targeting the wrong person: "Their quarrel is with the *Times'* editors, not me."

The hack's effect was also hotly debated. Markoff thought the hack had the potential to do "tremendous damage" to Kevin. If Kevin's defenders wanted to make the claim that Kevin and people like him are "harmlessly wandering through cyberspace," Markoff said, an event like this was the "clearest example to contradict that." Emmanuel Goldstein, editor of *2600,* saw things differently. "It's not what I would have done," Goldstein said, "but it got the story out. It is a story that has been suppressed for so long." The popular sentiment among hackers is that the coverage of the Mitnick case hyped his arrest and capture, referring to him as the "Internet's Most Wanted," as a "cyberthief," and in some cases as a "terrorist," but paid little or no attention to issues of Mitnick's pretrial incarceration, to the denial of his right to a bail hearing, or to the fact that the government had failed to provide Mitnick with access to the evidence to be presented against him.

Even Markoff, who insists that he played no part in putting Kevin in jail, indicated that he had "a lot of sympathy for Kevin," acknowledging that Mitnick was in a "difficult situation" and was faced with a "grim set of alternatives," but he rejected the notion that anyone but Mitnick himself was responsible for his situation: "Kevin made himself what he is."

The official statement from Mitnick's attorney was just as succinct: "Kevin Mitnick appreciates the support and good wishes of those who speak out against his continued state of incarceration for years without bail. However, he does not encourage any individuals to engage in hacking pranks on his behalf. Kevin believes other avenues exist that can be more beneficial to his circumstances," and he directed supporters to the Mitnick Web site at www.kevinmitnick.com.

The hack of the *New York Times* Web page did demonstrate a number of things. First, and most important, hackers were becoming activists. The hack of the *Times* was not just a prank to show the hackers' skills or for bragging rights; they had a message. Second, the movement that was unifying hackers was harking back to their early roots in the underground. In New York and Los Angeles, groups of hackers had held protests outside of Miramax offices to protest the filming of *Takedown* (the film based on Markoff and Shimomura's book); they had created an activist culture; and they had been *organizing*. The Kevin Mitnick mailing list was filled with all sorts of ideas, from door-to-door canvassing to flyer distribution at malls to making and selling "Free Kevin" mouse pads. Hackers were even willing to spend time outside of NBC studios in New York holding "Free Kevin" signs in the hopes that they would get air time on the *Today* show. The protests over the initial screenplay (which had been leaked to the hacker underground) resulted in a series of new scripts which corrected factual errors and resulted in a script which was much more sympathetic to Mitnick than earlier versions had been.

The legacy of Mitnick's case will be twofold. First, even though Mitnick's case never went to trail, a number of legal issues were confronted for the first time, giving a taste of what is to come in the future. Second, the "Free Kevin" movement, which his supporters vow will continue until he is totally free, taught hackers how to organize and how to create a movement that intervened politically, socially, and culturally over issues of law, justice, and representation. While Mitnick's case tested the boundaries of legal issues in court and started a hacker movement that may well continue on, another hacker, arrested at almost the same time, was fighting a different set of battles in federal court on appeal.

Minor Threat's Major Sentence

Chris Lamprecht steps up to a small computer terminal, punches in his ID number, and receives his current account balance from the commissary computer. As an inmate at the Federal Correctional Institution in Bastrop, Texas, Chris is allowed to check his balance using the prison's computer network. What is interesting about Chris's case is that such an act, using a networked computer, will be illegal for him once he is released from prison. According to his supervised release conditions, Chris, better know to hackers as "Minor Threat," is the first person in the United States to be banned from utilizing a computer network, including the Internet.

This ban is even more interesting because Chris is serving time for an offense entirely unrelated to computers, the Internet, or hacking. In fact, the word "Internet" was not even mentioned in Chris's case until the very last moments of sentencing, when the judge announced the conditions of the supervised release. In 1995, Chris was sentenced for a number of crimes, to which he plead guilty, involving the theft and sale of Southwestern Bell circuit boards. For those crimes, Chris certainly deserved punishment, including spending time right where he is — in federal prison. The problem with his sentencing was that Chris was also sentenced for being a hacker, something for which he has never been charged, tried, or prosecuted.

This was not the first time that a hacker had been prohibited from using a computer as a condition of supervised release. Similar penalties had been meted out in other hacker cases. Both Mitnick and Poulsen had restrictions placed on their computer usage as conditions of their release. What makes Chris's case different is both that he was banned specifically from the Internet and that his case was entirely unrelated to hacking or computers. In essence, Chris has been banned from the Internet for *being a hacker,* not for anything he has done or because his hacking in some way violated the law.

The information about Chris (Minor Threat) that caused him the most trouble came from two sources, his PSR (presentencing report) and tapes of a phone call made while he was in prison. In the case of the PSR, it had been discovered that Chris was a hacker of some stature in the computer underground. He had recently, in fact, been pro-philed in *Phrack,* where he espoused a philosophy of noncooperation with authorities, particularly turning in friends and fellow hackers. This became prima facie evidence of Chris's noncooperation, even though he had in actuality cooperated with authorities as a condition of his plea, providing them with information and evidence of his own crimes that they would have never discovered or had any reason to suspect. Chris's PSR had little relevance to his case, but instead focused on the fact that he was a hacker, and that would, ultimately, be the thing for which Chris was sentenced.

The second element, the recorded conversation, was even more damaging. A friend of Chris's, during a phone call to him, suggested a form of electronic retaliation against the police involved in his arrest. Chris rejected the idea, indicating that he didn't believe that such action was appropriate. The only problem was that the tape, which was never provided to the defense, was never played in court. Instead, the prosecution had a jail official testify to what the conversation was about. While the conversation *was* about retaliation, the fact that Chris opposed such action was omitted, leaving the judge with the impression that Lamprecht had, in fact, suggested, rather than rejected, the idea.

Based on this information, the judge's response made sense: this was a known hacker talking about retaliating against those who arrested and prosecuted him. (And, for all the judge knew, there might also be retaliation against the person who sentenced him.) But Lamprecht was not only banned from the Internet — he was also prohibited from serving as a "computer programmer, troubleshooter, or installer," the three jobs that he held before his arrest. This was in spite of the fact that his former employer offered to hire him back following his release. So why might a judge make such a decision? After the sentence and the conditions of his supervised release were announced, Chris was given an opportunity to respond. His remarks were short: "I mean, computers are my life." To this the judge responded, "I understand that. And that's why I put these conditions in, if you want to know the truth."

It is a long way from selling stolen goods to being banned from the Internet and your profession of choice. The move, as the judge's comments reveal, seems to be purely punitive. Punishment, however, is not the goal of supervised release. The goal of supervised release is to reintegrate the convict into society and make him or her a productive member of that society. Banning someone from the area where they are *most likely to be productive* seems counterintuitive at best. It also promises to make it very difficult for Chris to finish his degree in computer science at the University of Texas. Other hackers, denied the opportunity to make a living at what they did best and enjoyed the most, have often returned to hacking and occasionally did so with raised stakes. It is almost as if courts are intentionally working to turn hackers into what authorities fear most. Making it more difficult for hackers to take on legitimate jobs and turn their hobbies and obsessions into productive (even lucrative) careers is a recipe for disaster. By prohibiting hackers from using computers once released from prison, the judicial system is cutting off their only means for "going straight." In fact, most hackers who stay in the scene after college usually end up working as programmers, security consultants, or running their own systems.

The case is particularly acute in Lamprecht's situation. Chris is already a talented programmer. In the early 1990s, he wrote a program called ToneLoc, a phone dialing program that was modeled on the program Matthew Broderick uses in the movie *WarGames* to find open modem lines in telephone exchanges. The program was sophisticated enough to be embraced by both hackers and security experts, many of whom Chris helped to install and test the program, looking for security holes in their own systems. Later, two of these security experts, both from government agencies, were prohibited from testifying as character witness at Chris's trial: one explicitly, the other by being told he could not use government stationery to write a letter on Chris's behalf.

Chris, since owning his first computer, has taught himself to program in BASIC, Assembly, C, C++, and, since he has been in prison, Java. If allowed to work with computers again, he estimates that it will take him six months to get up to speed in Java to the point where he will be able to begin developing software. After his release, he wants to finish his degree and get a job programming — his true devotion. As things stand right now, none of those options is possible.

These circumstances raise serious issues that transcend the scope of Lamprecht's case in particular. In a recent decision, the Supreme Court held that the Internet is the most democratic form of expression and deserves the highest degree of protection. It is, in essence, the medium of free speech for the twenty-first century. To ban someone from the Internet, for an offense related neither to computers nor to the Internet itself, is at best punitive and at worst unconstitutional.

But, perhaps more important, Chris's sentence reveals how deeply embedded is the fear of hackers in the American judicial system. In this case, Chris pleaded guilty to one thing and was sentenced for something completely different and unrelated to his crime. In short, Lamprecht was found guilty of stealing (in several forms) and was (and should have been) punished for those crimes, but he was sentenced for being a hacker. What we need to question is how easily the figure of the hacker is transformed into a criminal, even when, as a hacker, the person has done nothing demonstrably wrong.

Most recently, Lamprecht got a rude awakening when he appeared in court in 1999. Chris, who has been in prison since 1995, appeared in court to argue that the government had breached its 1995 plea agreement, which led to Lamprecht's initial sentencing. What he wasn't prepared for was a government response to a second brief he had filed, which challenges the conditions of his supervised release.

As a result of his initial sentencing, Lamprecht will have as a condition of his supervised release a restriction that prohibits him from "utilizing any computer network," including the Internet, effectively making him the first person to be banned from the Internet.

In a brief filed the day before the hearing, the prosecutors responded to Lamprecht's motion to have those restrictions lessened, a tactic that caught Lamprecht and his attorney, Robert Kuhn, off-guard. "We were ambushed," Lamprecht said. "We had no notification and we were not ready to rebut their claims."

Since his incarceration, Michele Wood, Lamprecht's mother, has been maintaining a Web page that has provided information about Lamprecht's case and informed people about his "Internet ban." This Web page was at the center of the controversy around his 1999 court appearance. "They talked about me having a Web page like it was a horrible thing," Lamprecht said. Upon recommendation of his attorney, Lamprecht has decided to have the Web page taken down. "I guess the government has silenced me," he said. "I didn't think that this is how the first amendment was supposed to work."

Prosecutors see things differently, in large part stemming from an earlier incident where the underground hacker journal *Phrack* published the name and social security number of an IRS agent who had testified against Lamprecht. According to prosecutors, the agent suffered numerous incidents of harassment, including having his credit rating ruined. Lamprecht commented that "publishing his name in *Phrack* was the wrong thing to do. I'm sorry I ever did it," but he doesn't believe that it justifies an Internet ban. "If *Phrack* had been mailed, would they have banned me from using the mail? Of course not." The incident has left federal judge Sam Sparks and U.S. attorneys concerned about similar retaliations.

Prosecutors accused Lamprecht of running his Web page from prison, a claim he emphatically denies. The page, he says, was run and maintained by his mother for the sole purpose of educating people about his case. What is clear is that the Web page has generated media attention about Lamprecht's Internet ban, prompting several news stories and TV interviews with the jailed hacker.

Because of the last-minute nature of the government's filing, Lamprecht was not able to produce any witnesses on his behalf, nor was he able to testify, citing threats from prosecutors that he would be examined about alleged crimes not covered under his plea agreement if he were to take the stand on his own behalf. Although the court did not demand that Lamprecht remove his Web page, Lamprecht felt pressured. "I finally decided to take the page down, so I might be able to have a chance of using the Net when I get out next year."

Overall, Lamprecht's concerns are practical. When he is released from prison, he plans to continue his studies at the University of Texas, where he was majoring in computer science. The Internet ban, he fears, will make completing his degree next to impossible. Lamprecht's case continues, and, recently, he won back his right to direct appeal, something he had given up as a condition of his initial plea bargain. Regarding the Internet ban, Lamprecht said, "I'm still going to fight like hell. I'll just have to do it without a Web page."

The issues that Mitnick's and Lamprecht's cases raise are new legal, social, and cultural matters that will need to be faced in the coming years. Hacking is changing as fast as the technology that accompanies it. The issues that remain, however, will always be ones that focus primarily on human relationships and cultural attitudes toward technology, change, and difference.

By tracing out hacker culture, from its origins in the 1950s and 1960s through the various transformations it has taken in the 1980s and 1990s, this work has illustrated the complex ways in which technology has played a pivotal role in the formulation of the hacker underground and in the public, popular, and legal representation of it. Marking such transformations not only provides a sense of where hacker culture has come from but also comments on the role of technology in mainstream culture and illustrates the ways in which technology has been woven into the fabric of American society. Over the next decade, we can expect to see changes in the roles that hackers take on, the manner in which they negotiate their identity, and the ways in which they inform culture about the role of technology in the practice of everyday life.

Notes

Introduction

1. For an example of the debate over the nature of the term "hacker," see Paul Taylor, *Hackers: Crime in the Digital Sublime* (London: Routledge, 1999), 13–15.

2. Andrew Ross, "Hacking Away at the Counterculture," in *Technoculture*, ed. Constance Penley and Andrew Ross (Minneapolis: University of Minnesota Press, 1991), 121; Slavoj Zizek, "From Virtual Reality to the Virtualization of Reality," in *Electronic Culture: Technology and Virtual Representation*, ed. Timothy Druckrey (New York: Aperture, 1996), 293; Allucquere Rosanne Stone, *The War of Desire and Technology at the Close of the Mechanical Age* (Cambridge, Mass.: MIT Press, 1996). For more extended accounts, see Sherry Turkle, *The Second Self: Computers and the Human Spirit* (New York: Simon and Schuster, 1984), especially chapter 6.

3. William Gibson, *Neuromancer* (New York: Ace Books: 1983).

4. Thomas M. Disch, *The Dreams Our Stuff Is Made Of: How Science Fiction Conquered the World* (New York: Free Press, 1998), 220.

5. Bruce Sterling, personal interview, May 14, 1998.

6. A similar theme is explored in relation to skateboard culture and the reconstruction of space in Michael Nevin Willard, "Seance, Tricknowlogy, Skateboarding, and the Space of Youth," in *Generations of Youth: Youth Cultures and History in Twentieth Century America*, ed. Joe Austin and Michael Nevin Willard (New York: NYU Press, 1998), 327–46.

7. E. Anthony Rotundo, "Boy Culture," in *The Children's Culture Reader*, ed. Henry Jenkins (New York: NYU Press, 1998), 349.

8. Scott Bukatman, *Terminal Identity: The Virtual Subject in Postmodern Science Fiction* (Durham, N.C.: Duke University Press, 1993), 18.

9. Rotundo, "Boy Culture," 347–48.

10. Ibid., 349. For other discussions of boy culture, codes, and aggression, see William Pollack, *Real Boys: Rescuing Our Sons from the Myths of Boyhood* (New York: Random House, 1998); and Harvey Schwartz, "Reflections on a Cold War Boyhood," in *Boyhood: Growing Up Male* (Madison: University of Wisconsin Press, 1993), 165–75.

11. See, for example, discussions of the production of youth culture and gendered identity in underground "zines" in Stephen Duncombe, "Let's All Be Alienated Together: Zines and the Making of an Underground Community," and Willard, "Seance," both in *Generations of Youth*.

12. See, for example, Turkle, *The Second Self*. Turkle's discussion of hackers

(primarily old-school hackers) focuses on notions of control and mastery as characteristic of the hacker personality.

13. Of particular note is the manner in which literary theory has been used to better understand and account for the postmodern nature of technology. For example, see Mark Poster, "Theorizing Virtual Reality: Baudrillard and Derrida," in *Cyberspace Textuality: Computer Technology and Literary Theory*, ed. Marie-Laure Ryan (Bloomington: Indiana University Press, 1999); Poster, *The Mode of Information: Poststructuralism and Social Context* (Chicago: University of Chicago Press, 1990); N. Katherine Hayles, *Chaos Bound: Orderly Disorder in Contemporary Literature* (Ithaca, N.Y.: Cornell University Press, 1990); Hayles, "Virtual Bodies and Flickering Signifiers," in *Electronic Culture: Technology and Virtual Representation*, ed. Timothy Druckrey (New York: Aperture, 1996); Hayles, "Text out of Context: Situating Postmodernism within an Information Society," *Discourse* 9 (1987): 24–36; Janet Murray, *Hamlet on the Holodeck: The Future of Narrative in Cyberspace* (Cambridge, Mass.: MIT Press, 1997); Anne Balsamo, *Technologies of the Gendered Body: Reading Cyborg Women* (Durham, N.C.: Duke University Press, 1996). For extended discussion of postmodernity and fiction, see Steven Shaviro, *Doom Patrols* (New York: Serpent's Tail, 1997); and Brian McHale, *Postmodern Fiction* (New York: Methuen, 1987).

14. For an extended reflection on technology and postmodernity, see Steven Shaviro's "Theoretical Fiction," in *Doom Patrols*. Shaviro discusses the manner in which technology needs to be understood as a kind of postmodern fiction.

15. This idea is developed at length by Nicholas Negroponte in *Being Digital* (New York: Knopf, 1985).

16. François Lyotard, *The Postmodern Condition*, trans. Geoff Bennington and Brian Massumi (Minneapolis: University of Minnesota Press, 1984). For an additional examination of the relationship between critical theory and technology, see George Landow, *Hypertext: The Convergence of Contemporary Critical Theory and Technology* (Baltimore: Johns Hopkins University Press, 1992).

17. For more extensive discussions of the manner in which self and identity are impacted by technology, see Sherry Turkle, *Life on the Screen: Identity in the Age of the Internet* (New York: Simon and Schuster, 1995); and Jay David Bolter and Richard Grusin's discussion of the impact of remediation on notions of self in *Remediation: Understanding New Media* (Cambridge, Mass.: MIT Press, 1999).

18. Michel Foucault, "Two Lectures," in *Power/Knowledge* (New York: Pantheon, 1977), 81.

19. Ibid., 81–82.

20. Ibid., 83.

1. Hacking Culture

1. See Scott Bukatman, "Gibson's Typewriter," *South Atlantic Quarterly* (fall 1993): 627.

2. Paul Mungo and Bryan Clough, *Approaching Zero: The Extraordinary*

World of Hackers, Phreakers, Virus Writers, and Keyboard Criminals (New York: Random House, 1992), xvii, xviii.

3. Joe Chidley, "Cracking the Net," *Maclean's Magazine* (May 22, 1995): 54.

4. Amy Harmon, "Computer World Expects Devil of a Time with Satan Program," *Los Angeles Times*, March 1, 1995.

5. "Raising Hell on the Internet," *Oakland Tribune* and *San Jose Mercury News*, March 1, 1995.

6. SAGE (System Administrators Guild) Advisory, "What's All This about SATAN?" n.d.; available at http://www.vsenix.org/sage.

7. A similar argument is made by Arnold Pacey regarding the technical, organizational, and cultural aspects of technology; see his *Cultures of Technology* (Cambridge, Mass.: MIT Press, 1985).

8. Steven Levy, *Hackers: Heroes of the Computer Revolution* (New York: Anchor/Doubleday Press, 1984), 27–33.

9. Ibid., 91.

10. Steve Mizrach (aka Seeker1), "Old Hackers, New Hackers: What's the Difference?"; electronic publication available at http://www.eff.org.

11. Levy, *Hackers*, 27.

12. Ibid., 159.

13. For an extensive examination of the relationship between technological and military industrialization, see Manuel de Landa, *War in the Age of Intelligent Machines* (Cambridge, Mass.: Zone Books, 1992).

14. Levy, *Hackers*, 143.

15. Ibid., 27.

16. Bruce Sterling, *The Hacker Crackdown* (New York: Bantam, 1992), 43.

17. Ibid.

18. Katie Hafner and John Markoff, *Cyberpunk: Outlaws and Hackers on the Computer Frontier* (New York: Simon and Schuster, 1991), 20.

19. Levy, *Hackers*, 123.

20. The original article was titled "Secrets of the Black Box," by Ron Rosenbaum. It appeared in a 1971 issue of *Esquire*. It is reprinted in Rosenbaum's *Rebirth of the Salesman: Tales of the Song and Dance 70's* (New York: Doubleday, 1979).

21. Levy, *Hackers*, 242.

22. Ibid., 266.

23. Philip K. Dick, *Ubik* (New York: Vintage, 1969), 3.

24. William Gibson, *Neuromancer* (New York: Ace Books, 1984).

25. Ibid., 43.

26. Derrida makes much of this point in his *Given Time I: Counterfeit Money*, trans. Peggy Kamuf (Chicago: University of Chicago Press, 1992). See particularly his discussion of the relationship between the notions of *oikos* (home) and *nomos* (law) in the construction of economy (*oikonomia*) (p. 6). These notions are explored through the thematics of "the gift" and of "time" in the remainder of that work.

27. Anarchy and AoC, "Declaration of Digital Independence."

28. Ibid.

29. Paul Virilio, *The Art of the Motor*, trans. Julie Rose (Minneapolis: University of Minnesota Press, 1995), 3.

30. Anarchy and AoC, "Declaration of Digital Independence."

31. For an extensive discussion of technology and Cold War politics, see Paul Edwards, *The Closed World: Computers and the Politics of Discourse in Cold War America* (Cambridge, Mass.: MIT Press, 1996), 325–27.

32. Sterling, *Hacker Crackdown*, 84–85.

33. Taran King, "Phrack Pro-Phile XXVIII," *Phrack* 3, no. 28, file 2 (February 1, 1986).

34. Crimson Death, "Phrack Pro-Phile XXXIII," *Phrack* 3, no. 33, file 2 (February 1, 1986).

35. The story of these West German hackers is chronicled by Hafner and Markoff in *Cyberpunk*, 139–250. The influence of *WarGames* is detailed on p. 190.

36. The entire Robert Morris story is documented in ibid., 253–341. On. p. 302, Hafner and Markoff document the problems with the worm's propagation.

37. Ibid., 261.

38. See, for example, the entire issue of *Communications of the ACM* 32, no. 6 (June 1989), which is devoted to an analysis of Morris's worm program, and Eugene H. Spafford, "The Internet Worm Program: An Analysis," *Computer Communication Review* 19, no. 1 (January 1989): 17–57.

39. Hafner and Markoff, *Cyberpunk*, 321.

40. Mizrach, "Old Hackers."

41. John Perry Barlow, "Crime and Puzzlement" (1990), electronic publication posted to The Well.

42. Ibid., 1.

43. Clifford Stoll, *The Cuckoo's Egg: Tracking a Spy through the Maze of Computer Espionage* (New York: Pocket Books, 1989), 354.

44. Mizrach, "Old Hackers," 1.

45. Ibid., 2.

46. Barlow, "Crime and Puzzlement," 5.

47. Jim Thomas, "Hollywood Hacker or Media Hype?" *Computer Underground Digest* 3, no. 3.09 (March 19, 1991).

48. Barlow, "Crime and Puzzlement," 5.

49. This reference to the 1961 Rose Bowl between Washington and Minnesota is cited as a prototypical hack for the 1960s hacker both in Levy, *Hackers*, and in Hafner and Markoff, *Cyberpunk*, 11.

50. Hafner and Markoff, *Cyberpunk*, 336.

51. Virilio, *Art of the Motor*, 52–53, 33.

52. Levy, *Hackers*, 224.

53. Quoted in ibid., 224–25.

54. Ibid., 225.

55. Ibid., 342.

56. This theme was developed by Jacques Derrida in his public lecture series at the University of California, Irvine, in 1997, under the title "Hospitality/Hostility."

57. N. Derek Arnold, *UNIX Security: A Practical Tutorial* (New York: McGraw Hill, 1993), 196.

58. Olaf Kirch, *Linux Network Administrator's Guide* (Seattle: Specialized Systems Consultants, 1994), 268.

2. Hacking as the Performance of Technology

1. E. Anthony Rotundo, "Boy Culture," in *The Children's Culture Reader,* ed. Henry Jenkins (New York: NYU Press, 1998), 351.

2. My argument is similar to arguments developed by Turkle in her discussion of opaque and transparent technologies as well as Bolter and Grusin's discussion of transparent immediacy. The primary difference with my argument is the focus on the relational spaces opened up by such transformations, rather than the effects of the technology as an interface for individual users. See Sherry Turkle, *Life on the Screen: Identity in the Age of the Internet* (New York: Simon and Schuster, 1995); and Jay David Bolter and Richard Grusin, *Remediation: Understanding New Media* (Cambridge, Mass.: MIT Press, 1999).

3. Paul Edwards contextualizes the film in his book *The Closed World: Computers and the Politics of Discourse in Cold War America* (Cambridge, Mass.: MIT Press, 1996), 325–27.

4. Interestingly, "cyberspace" is a term that emerged from cyberpunk literature in the early 1980s and was popularized by John Perry Barlow (one of the inventors of The Well). Bruce Sterling attributes the popularization of Gibson's terminology to Barlow's usage of it, claiming that it was the term, "as Barlow employed it, [that] struck a useful chord, and this concept of cyberspace was picked up by *Time, Scientific American,* computer police, hackers, and even constitutional scholars" (Bruce Sterling, *The Hacker Crackdown: Law and Disorder on the Electronic Frontier* [New York: Bantam, 1992], 236).

5. Gareth Branwyn, "Hackers: Heroes or Villains?" introduction to Dennis Fiery (aka The Knightmare), *Secrets of a Superhacker* (Port Townsend, Wash.: Loompanics, 1994), 1.

6. Ibid.

7. For an analysis of the question of identity and cyberpunk literature, see Scott Bukatman, *Terminal Identities: The Virtual Subject in Postmodern Science Fiction* (Durham, N.C.: Duke University Press, 1993).

8. Joe Chidley, "Cracking the Net," *Maclean's Magazine* (May 22, 1995): 54.

9. Witness, in particular, the continuing effort to define hacker codes of ethics. As Fiery notes, "Many hackers and non-hackers have given their versions of the 'Hacker's Ethic.' The versions are all pretty much the same. What's different is the degree to which the ethic is followed" (Fiery, *Secrets of a Superhacker,* 161). What such "codes" point to is precisely a set of technological questions: How will I use these tools? What is a useful, productive, or good end? Almost always, such codes articulate a set of intentions that include some permutation on the idea of "doing no harm."

10. Martin Heidegger, "The Question concerning Technology," in *Basic*

Writings, ed. David Farrell Krell, trans. William Lovitt (New York: Harper and Row, 1977), 287.

11. Steven Levy, *Hackers: Heroes of the Computer Revolution* (New York: Anchor Press/Doubleday, 1984).

12. Both of these groups are chronicled in Sterling's *Hacker Crackdown.*

13. For example, the underground computer magazine *Phrack* has served the roles of both disseminating technical information about hacking and commenting on the culture that defines the hacker community. As Chris Goggans (aka Erik Bloodaxe), a former editor of *Phrack,* states: "[*Phrack*] has always tried to paint a picture of the social aspects of the computer underground rather than focus entirely on the technical issues. So in a way it adds a lot of color to what is going on" ("Interview with Chris Goggans at Pumpcon, 1993," *Gray Areas* 3, no. 2 [fall 1994]: 27).

14. In the past decade, there has been a proliferation of books chronicling the exploits of in/famous hackers such as Robert Morris, Kevin Mitnick, Kevin Poulsen, Pengo, and groups such as the Masters of Deception and the Legion of Doom. These accounts have almost exclusively been journalistic in style, documenting the "downfall" of "computer criminals."

15. The notion of the "elite" class of hackers is borrowed, again, from William Gibson's *Neuromancer* (New York: Ace Books, 1984).

16. Heidegger, "Question," 287.

17. Walter Benjamin, "The Task of the Translator," trans. Harry Zohn, in *Illuminations,* ed. Hannah Arendt (New York: Schocken Books, 1969), 70.

18. Ibid., 75.

19. I refer to Plato's analysis of Lysias's speech in *Phaedrus,* which, initially presented in writing, becomes the object of Socrates' critique. Writing, Plato argues, "when it is ill-treated and unfairly abused always needs its parent to come to its help, being unable to defend or help itself" (275e). Hence, Socrates' critique of Lysias enacts the defenselessness of writing itself.

20. Michel Foucault, "What Is an Author?" in *The Foucault Reader,* ed. Paul Rabinow (New York: Pantheon Books, 1984), 105, 114, 113.

21. Ibid., 119.

22. Sterling, *Hacker Crackdown,* 45.

23. "Interview with Chris Goggans at Pumpcon, 1993," 32.

24. Foucault, "What Is an Author," 116.

25. The *Hacker's Jargon Dictionary* is a long-standing text file that has continually been expanded and updated over the years to provide a standard reference for terms that are important to computer and underground culture but are not included in standard dictionaries or references. A published version is available under the title *The New Hacker's Dictionary* (Cambridge, Mass.: MIT Press, 1993).

26. As Fiery describes it, "Social engineering is the act of talking to a system user, pretending that you are also a legal user of the system, and in the course of the conversation, manipulating the discussion so that the user reveals passwords or other good stuff" (Fiery, *Secrets of a Superhacker,* 24).

27. Ibid., 19.

28. Ibid., 50.

29. Ibid.

30. Ibid., 54.

31. Ibid.

32. Robert Cringely, *Accidental Empires: How the Boys of Silicon Valley Make Their Millions, Battle Foreign Competition, and Still Can't Get a Date* (New York: Harper Business, 1996), 61.

33. Ibid., 62.

34. For an extensive discussion of the nature of ubiquitous computing, see Jay David Bolter and Richard Grusin, *Remediation: Understanding New Media* (Cambridge, Mass.: MIT Press, 1999), 213–19.

35. Most systems have implemented a security system called "shadowing," which hides the encrypted portion of the password file, making the "passwd" file worthless to the hacker. Such a system significantly improved UNIX system security.

36. Fiery, *Secrets of a Superhacker*, 40.

37. Ibid., 48

38. Donna Haraway, "Interview with Donna Haraway," in *Technoculture*, ed. Andrew Ross and Constance Penley (Minneapolis: University of Minnesota Press, 1991), 6.

39. Haraway's notion of cyborg identity has been taken up and widely discussed from a number of vantage points. See, for example, Chris Hables Gray et al., *The Cyborg Handbook* (New York: Routledge, 1996). My intention is not to argue that hackers are or are not in fact cyborgs, but instead to situate the notion of a hybrid/deconstructive identity position within the discourse of technology and culture.

40. The Mentor, "The Conscience of a Hacker," *Phrack* 1, no. 7, file 3 (1985). Originally published in *Phrack* in 1985, the essay has taken on a life of its own. In most cases, it is still attributed to The Mentor, but the title is often changed to "The Hacker Manifesto" or, in one case, "Mentor's Last Words."

3. Hacking in the 1990s

1. Arjun Appadurai, *Modernity at Large: Cultural Dimensions of Globalization* (Minneapolis: University of Minnesota Press, 1996), 34.

2. Linus Torvalds, comp.os.minix, August 1991; reprinted in *The Official Red Hat Linux Installation Guide 5.1* (Research Triangle Park, N.C.: Red Hat Software, 1998), xvii.

3. Ibid.

4. Ibid.

5. *Microsoft Windows User's Guide* (Redmond, Wash.: Microsoft Corporation, 1990), ix.

6. *Microsoft Windows Operating System Version 3.1 User's Guide* (Redmond, Wash.: Microsoft Corporation, 1993), xiii.

7. Ibid.

8. *Introducing Microsoft Windows 95* (Redmond, Wash.: Microsoft Corporation, 1995), v.

9. I borrow these terms from Sherry Turkle, *Life on the Screen: Identity in the Age of the Internet* (New York: Simon and Schuster, 1995).

10. "Phrack Pro-Phile III Featuring: User Groups and Clubs," *Phrack* 1, no. 6, file 2 (June 10, 1986).

11. "Phrack Pro-Phile IV," *Phrack* 1, no. 7, file 2 (June 28, 1986).

12. "PWN SummerCon '87 Quicknotes" *Phrack* 1, no. 14, file 1 of 9 (July 28, 1987).

13. Deth Vegetable, interview, August 12, 1998.

14. Oxblood Ruffin, interview, October 1, 1997.

15. Ibid.

16. Blondie Wong, director of the Hong Kong Blondes, interview, in Oxblood Ruffin, "The Longer March," *cDc* (online journal), file 356 (June 15, 1998).

17. Microsoft market bulletin, "Microsoft's Response to the Cult of the Dead Cow's 'Back Orifice' Tool," August 4, 1998.

18. Deth Vegetable, interview, August 12, 1998.

19. Ibid.

20. Ibid.

21. Ibid.

22. Microsoft security bulletin (MS98–010), "Information on the 'Back Orifice' Program," August 12, 1998.

23. Ibid.

24. Ibid.

25. Cult of the Dead Cow, "Morality Alert," August 10, 1998.

26. Wong, in Oxblood Ruffin, "The Longer March."

27. Ibid.

28. Ibid.

29. Ibid.

30. Microsoft Corporation, "Response to Security Issues Raised by the L0phtcrack Tool," April 18, 1997.

31. Microsoft Corporation, "Clarification on the L0phtcrack 2.0 Tool," February 14, 1998.

32. L0phtCrack 2.0 manual.

33. "Mass High Tech Article on AIP 'Hacker' Talk," letter to Mass High Tech, posted on http://www.L0pht.com/mht-rebuttal.html.

4. Representing Hacker Culture

1. Bruce Sterling, *The Hacker Crackdown: Law and Disorder on the Electronic Frontier* (New York: Bantam, 1992), 44–45.

2. Ibid., 65.

3. *2600: The Hacker Quarterly* 15, no. 1 (spring 1998): 23.

4. Sterling, *Hacker Crackdown*, 85.

5. Ibid., 86.

6. BellSouth's E911 document, cited in ibid., 120.

7. Sterling, *Hacker Crackdown*, 112–42, 240–72; and Paul Mungo and Bryan Clough, *Approaching Zero: The Extraordinary Underworld of Hackers,*

Phreakers, Virus Writers, and Keyboard Criminals (New York: Random House, 1992).

8. Mungo and Clough, *Approaching Zero,* 211.

9. Sterling, *Hacker Crackdown,* 124.

10. Ibid.

11. Ibid., 126.

12. John Perry Barlow, "Crime and Puzzlement" (1990), electronic publication posted to The Well.

13. Mungo and Clough, *Approaching Zero,* 220.

14. These costs are documented in Sterling's account in *Hacker Crackdown,* 246–47.

15. *Phrack* copyright, 1996.

16. "Interview with Chris Goggans at Pumpcon, 1993," *Gray Areas* (fall 1994): 27–50.

17. For extensive discussions of issues of privacy and copyright, see Philip E. Agre and Marc Rotenberg, *Technology and Privacy: The New Landscape* (Cambridge, Mass.: MIT Press, 1998) and Whitfield Diffie and Susan Landau, *Privacy on the Line* (Cambridge, Mass.: MIT Press, 1998).

18. *Phrack* registration file.

19. "Interview with Chris Goggans," 28.

20. Ibid.

21. Ibid.

22. "Phrack Pro-Phile II (Broadway Hacker)," *Phrack* 2, no. 5, file 1.

23. "Phrack Pro-Phile XXII (Karl Marx)," *Phrack* 2, no. 22, file 2.

24. "Phrack Pro-Phile XXVIII (Eric Bloodaxe)," *Phrack* 3, no. 28, file 2.

25. Ibid.

26. "Phrack Pro-Phile XXIII (The Mentor)," *Phrack* 2, no. 23, file 2.

27. "Phrack Pro-Phile XXII (Karl Marx)."

28. Michel Foucault, "What Is an Author?" in *The Foucault Reader,* ed. Paul Rabinow (New York: Pantheon Books, 1984) 115–16.

29. Ibid., 121.

30. Ibid., 123.

31. *Phrack* 1, no. 3, file 10.

32. *Phrack* 4, no. 40, file 12.

33. *Phrack* 10, no. 56, file 4.

5. (Not) Hackers

1. Dick Hebdige, *Subcultures: The Meaning of Style* (New York: Routledge, 1979), 3, 87.

2. Ibid., 92.

3. Sherry Turkle, *Life on the Screen: Identity in the Age of the Internet* (New York: Simon and Schuster, 1995), 57.

4. Ibid., 52.

5. Jay David Bolter and Richard Grusin, *Remediation: Understanding New Media* (Cambridge, Mass.: MIT Press, 1999), 5.

6. Hebdige, *Subcultures,* 95.

7. Ibid.
8. See, for example, Bruce Sterling, *The Hacker Crackdown: Law and Disorder on the Electronic Frontier* (New York: Bantam, 1992).
9. Hebdige, *Subcultures*, 103–4.
10. T. Hawkes, *Structuralism and Semiotics* (London: Methuen, 1977); cited in Hebdige, *Subcultures*, 103.
11. Hebdige, *Subcultures*, 107, 92–94.
12. Ibid., 97.
13. Ibid., 95–7.
14. *Hackers* Web site, http://www.mgm.com/hackers.

6. Technology and Punishment

1. Katie Hafner and John Markoff, *Cyberpunk: Outlaws and Hackers on the Computer Frontier* (New York: Simon and Schuster, 1991), 9.
2. Mitchell Kapor, "A Little Perspective, Please," *Forbes Magazine* (June 21, 1993).
3. *United States v. Edward Cummings*, no. 95–320, June 8, 1995.
4. Hafner and Markoff, *Cyberpunk*, 11.
5. Michel Foucault, *Discipline and Punish: The Birth of the Prison*, trans. Alan Sheridan (New York: Vintage, 1977).
6. Ibid.
7. Ibid.
8. Friedrich Nietzsche, *On the Genealogy of Morals*, trans. Walter Kaufman (New York: Viking, 1969), 2.12, 3.
9. Ibid., 2.3.
10. Ibid.
11. "Interview with Chris Goggans at Pumpcon, 1993," *Gray Areas* (fall 1994): 37.
12. See, for example, Title 18, United States Code, Section 1029(b)(2).
13. It also refers to items that allow the hacker to hide his or her identity, a point that will be taken up below in the discussion of panopticism.
14. This problem is also at the heart of the origins of and debate over the implementation of cryptographic systems. Cryptography can be read, in this context, as a way to negotiate these questions of identity in a more sophisticated and complex manner.
15. Michel Foucault, *The History of Sexuality*, vol. 1, trans. Robert Hurley (New York: Pantheon, 1978), 34.
16. "Interview with Chris Goggans."
17. On the term "cyberspace," see n. 4 for chap. 2, above.
18. William Gibson, *Neuromancer* (New York: Ace Books, 1984), 6.
19. While there are documented cases of hackers engaging in serious computer crime, more often, the efforts of law enforcement are aimed at cracking down on fairly innocuous behavior. For an instance of the former, see Clifford Stoll, *The Cuckoo's Egg: Tracking a Spy through the Maze of Computer Espionage* (New York: Pocket Books, 1989). For an analysis of the latter, see Bruce

Sterling, *The Hacker Crackdown: Law and Disorder on the Electronic Frontier* (New York: Bantam, 1992).

20. Hafner and Markoff, *Cyberpunk*, 342, 343.

21. Letter from Marc J. Stein, U.S. probation officer, to Kevin Lee Poulsen, May 22, 1996. Poulsen's saga continued after his release.

22. Kevin Poulsen, "Many Happy Returns" (http://www.kevinpoulsen.com).

23. John Markoff, "Cyberspace's Most Wanted: Hacker Eludes FBI Pursuit," *New York Times*, July 4, 1994.

24. Hafner and Markoff, *Cyberpunk*, 26.

25. See Markoff, "Cyberspace's Most Wanted."

26. Ibid.

27. Joshua Quittner, *Time*, February 27, 1994.

28. Hafner and Markoff, *Cyberpunk*, 343, 344.

29. On this point, particularly, see Jacques Derrida's essay "Plato's Pharmacy," in *Dissemination*, trans. Barbara Johnson (Chicago: University of Chicago Press, 1983).

30. Mike Godwin, in Douglas Fine, "Why Is Kevin Lee Poulsen Really in Jail?" Posting to The Well, 1995.

31. John Markoff, interview, January 19, 1995.

32. U.S. Marshals Service, NCIC entry number NIC/W721460021.

33. John Markoff, "A Most-Wanted Cyberthief Is Caught in His Own Web," *New York Times*, 15 February, 1995.

34. For a detailed account of Mitnick's deeds and misdeeds, see Hafner and Markoff, *Cyberpunk*, 13–138. Hafner and Markoff chronicle Mitnick's life as a "dark-side hacker," which has been considered by some (in particular Emmanuel Goldstein and Lewis DePayne) to be a misnomer. Advocates for Mitnick are quick to note that Mitnick's hacks, while occasionally malicious, were never intended to cause harm to the victims or produce any financial gain for Mitnick.

35. John Markoff, "A Most-Wanted Cyberthief."

36. Ibid.

37. Ibid.

38. Jonathan Littman, *The Fugitive Game: Online with Kevin Mitnick* (New York: Little Brown, 1997).

39. Tsutomu Shimomura and John Markoff, *Takedown: The Pursuit and Capture of Kevin Mitnick, America's Most Wanted Computer Outlaw — by the Man Who Did It* (New York: Hyperion, 1996), 308.

40. Ibid., 309.

41. Ibid., 310.

42. Charles Platt, *Anarchy Online: Net Crime* (New York: Blacksheep Books, 1996).

43. As LOD members are quick to point out, LOD was originally a group of phone phreaks. As phone phreaking evolved and began to involve computers, the group formed a "hacker" splinter group, who called themselves LOD/H, or Legion of Doom/Hackers, to distinguish themselves from the original phone-phreaking group.

44. E. Anthony Rotundo, "Boy Culture," in *The Children's Culture Reader*, ed. Henry Jenkins (New York: NYU Press, 1998), 349.

45. Ibid.
46. Lex Luthor, "The History of LOD/H," *LOD/H Technical Journal* 4, file 6 (May 1990).
47. "Interview with Chris Goggans," 34.
48. Luthor, "The History of LOD/H," 2.
49. "Interview with Chris Goggans," 34.
50. "Phrack Pro-Phile," *Phrack* 1, no. 6, file 2 (June 10, 1986).
51. "Interview with Chris Goggans," 35.
52. Michelle Slatalla and Joshua Quittner, *Masters of Deception: The Gang That Ruled Cyberspace* (New York: Harperperennial Library, 1996), inside book jacket.
53. Ibid., 139.
54. Ibid., 139, 142; emphasis added.
55. K. K. Campbell, "Bloodaxe Comes Out Swinging: *Phrack* editor Chris Goggans on *Masters of Deception*," *Eye Weekly: Toronto's Arts Newspaper* (September 7, 1995).
56. Chris Goggans, "The Real Master of Deception?" *Wired* 3.04 (April 1995).
57. Cited in Slatalla and Quittner, *Masters of Deception*, 206.
58. Ibid.
59. Northern California indictment (http://www.kevinpoulsen.com).
60. Kevin Poulsen, letter to Manuel L. Real, February 9, 1995.
61. Ibid.
62. Ibid.
63. See Douglas Fine, "Why Is Kevin Lee Poulsen Really in Jail?" Posting to The Well, 1995.
64. Northern California Indictment.
65. Jonathan Littman, *The Watchman: The Twisted Life and Crimes of Serial Hacker Kevin Poulsen* (New York: Little Brown, 1997).
66. Ibid., 244.
67. Ibid., 246.
68. Sigmund Freud, *Introductory Lectures on Psychoanalysis* (New York: W. W. Norton, 1966), 494–95.
69. Ibid., 214.

Epilogue

1. These reports are based entirely on conversations with the principal parties involved over a period of four years, including Kevin Mitnick, Ron Austin, Don Randolph, Greg Vinson, David Schindler, Richard Sherman, Chris Lamprecht, Michele Wood, Evian S. Sim, Robert Pitman, Kevin Poulsen, and Emmanuel Goldstein.

Index

Douglas Thomas is associate professor in the Annenberg School for Communication at the University of Southern California, Los Angeles.